Migration, Risk and Uncertainty

Migration is one of the driving forces of economic and social change in the modern world. It is both informed by risk and a generator of risk, whether for individuals, households, communities or societies. Although the relationship between migration and risk is widely acknowledged, it has long been neglected in academic research, with a few exceptions such as household diversification strategies. Instead, risk is assumed to be implicit in economic or social models, rather than being explicitly theorised or analysed. This book represents the first major review of these key relationships. It draws on a wide range of theories—from economics, psychology, sociology, anthropology and geography—and an equally broad range of empirical material, to provide a highly original overview.

Allan M. Williams is Professor of Tourism and Mobility Studies in the Faculty of Business, Economics and Law at the University of Surrey.

Vladimír Baláž is Research Professor with the Institute for Forecasting of the Slovak Academy of Sciences.

Routledge Studies in Human Geography

This series provides a forum for innovative, vibrant, and critical debate within Human Geography. Titles will reflect the wealth of research which is taking place in this diverse and ever-expanding field. Contributions will be drawn from the main sub-disciplines and from innovative areas of work which have no particular sub-disciplinary allegiances.

Published:

For a full list of titles in this series, please visit www.routledge.com

Migration, Risk and Uncertainty

Allan M. Williams and Vladimír Baláž

Routledge
Taylor & Francis Group

LONDON AND NEW YORK

First published 2015 by Routledge

2 Park Square, Milton Park, Abingdon, Oxfordshire OX14 4RN
711 Third Avenue, New York, NY 10017

*Routledge is an imprint of the Taylor & Francis Group,
an informa business*

First issued in paperback 2018

Library of Congress Cataloging-in-Publication Data
A catalog record for this book has been requested.

ISBN: 978-0-415-65952-9 (hbk)
ISBN: 978-1-138-54709-4 (pbk)

Typeset in Sabon
by Apex CoVantage, LLC

Contents

Preface

Risk and uncertainty are integral to migration—whether in terms of shaping, or being shaped by, migration. Leaving a familiar setting, supportive networks of friends and kin, and moving to a new country that you have less knowledge of, perhaps not knowing its language or culture, and having very few or no close friends there, necessarily involves risk and uncertainty. So too does non-migration: not only when faced with known severe economic or political challenges in your home area, but also the uncertainties of the future in any area, or the risk that you will have foregone a chance of greater economic success, higher welfare, or happiness if you had moved. In addition to these risks for the individuals, there are also the risks for their families—who move with them or who are left behind—and for the communities and countries of origin and destination. Even where researchers have addressed the issues of risk and uncertainty in migration, they have often done so from diverse theoretical perspectives that are not always made explicit, or understood in terms of their limitations as well as their contributions to unravelling this nexus of relationships.

It is surprising therefore that the engagement of migration researchers with risk and uncertainty has been highly uneven, and has often been implicit rather than explicit. In addressing that lacuna, this book is the first systematic attempt to provide an overview of migration, risk and uncertainty, one that is multidisciplinary and multi-scalar. It has grown out of the authors' earlier research on international migration and knowledge, and from their reflections on the limits to knowledge, and how the edges of knowledge blur into risk and uncertainty.

This is in essence a book about how we can theorise the role of risk and uncertainty in migration. Its primary focus is the arena of academic debate and research, but it addresses a topic of immense, and substantive, importance to migrants, communities and countries, and to policy makers. Open the newspapers, watch the television, or visit Internet-based news websites in most countries, and there is likely to be some discussion, or reporting, of the risks (usually perceived in negative terms) either experienced by migrants or consequent on migration. These themes also have a strong presence in policy discourses in many countries, and in politics also, although the language is often that of security, threats, and dangers, rather than explicitly of risk and uncertainty.

Given the ambitious scope of this book, and the broad theoretical fields that it seeks to engage with, a high degree of selectivity has been necessary in focussing on particular aspects of the relationships between migration, risk and uncertainty, and this is especially so given the diverse forms of migration. Nevertheless, the book does aim to advance understanding of the topic, especially through its exploration of the competing and complementary perspectives provided by different theories. We hope it will be a platform for future research and perhaps for the meeting of some of these different theoretical voices.

We are grateful for the comments we have received on our work when presenting it at conferences and workshops, and to the referees of migration and risk journals. And we also gratefully acknowledge the financial assistance provided by the Economic and Social Research Council for our initial work on knowledge and migration, and by the Leverhulme Trust and the Centre of Excellence of the Slovak Academy of Sciences (CESTA) (grant no. III/2/2011) for our work on the behavioural economics of the willingness of migrants to take risks.

Allan M. Williams
(University of Surrey, Guildford)

and

Vladimír Baláž
(Institute of Forecasting, Bratislava)

1 Introduction

MIGRATION AND RISK: THE CHALLENGES

Risk and uncertainty are intrinsic and contingent features of human activities, ranging from the trivial to the substantial and life-changing (Tulloch and Lupton 2003: 1; Mehta 2007: 3). Risk is especially evident, and inherent, in migration, whether in terms of individual decision making and outcomes, or the implications for households, communities and national states (Williams and Baláž 2012). It is manifested, although in different ways, in all forms of migration—whether as refugees or asylum-seeking, labour or lifestyle migrants. Despite the pervasiveness of risk issues in migration and migration policies, there has been surprisingly little explicit attempt to engage with and analyse these issues, with a few important exceptions. This book provides a multi-scalar, multidisciplinary review of how the role of risk and uncertainty in migration have been, and can be, conceptualised.

The need for greater theoretical clarity in the understanding of the role of risk in migration is underlined by the almost constant presence in newspapers, or internet news pages, of issues relating to what are perceived to be risks associated with migration. These may be risks for individuals, especially in terms of migrants seen to risk their lives in highly hazardous irregular crossings of borders, or the risks of exploitation or of failing to find jobs. Or it may be the risks to the sending countries from brain drain or from fluctuations in remittances, or the risks to the destination countries in the arenas of employment, health and crime. Such risks are real and 'out there', but they are also socially constructed, and stand at the confluence of powerful and competing discourses and narratives about the nature of migration, and the risks and uncertainties relating to rewards and costs to different parties.

There is, of course, nothing new about large scale migration (Chiswick and Hatton 2003), or about hotly contested discourses about the consequent risks (see Box 1.1). Riots, protests and discrimination against immigrants are written large throughout history, as illustrated by Richard Bean's controversial play *England People Very Nice* (first performance, National Theatre, London, February 2009), a voyage through the risks faced by successive groups of immigrants in London over a 400-year period.

However, fluctuations in the scale of migration shape the framing of discourses. For example, Flynn (2005: 466) considers that, in the UK, a growing sense of crisis was informed by a general intensification in mobility levels, as well as specific increases in immigration: a 'feature which marked the growing sense of crisis was the increase in the volume of international travel from the mid-1980s onwards. In 1985 the total number of people passing through UK ports of entry was in the region of 40 million. . . . Over the next ten years this figure doubled to reach 80 million by 1997/8, and over 90 million by the end of the millennium'. It is not only the scale of migration which has changed, but also how this is seen as emblematic of globalization (Held 2000), whereby an increasing array of places have become increasingly interdependent, linked by trade, migration and knowledge flows. Events in distant countries, whether economic, political or natural disasters, are seen to result in substantial, and often unanticipated, flows of people through global networks, with attendant risks for all those involved. In other words, although migration risks are most starkly manifested at the local scale in both destination and origin countries, they are understood as being at the nexus of the global and the local.

Box 1.1   The History of the World as a History of Migration

Alexander the Great is renowned for his military achievements, but his army was also accompanied by traders, artists and scientists. Conquered lands—Alexander founded 20 cities that bore his name—were colonised by veteran soldiers and immigrants from Greece. In this way, Alexander and his successors built the 'Hellenistic World', which ranged from Italy to India (Wood 2001). The Roman Empire also combined military expansion with colonisation. Roman migrants—soldiers, veterans and their families, as well as businessmen and administrators—colonised territories from the British Isles to Mesopotamia. Subsequently, massive population movements in the period 400–800 AD shaped the current ethnic map of Europe. Members of Germanic, Slavic, Berber/Arab and Magyar tribes founded large numbers of 'barbarian' kingdoms. Some of these states were forerunners of contemporary European countries (Noble and Goffart 2006).

Early colonization by the Spanish, Portuguese, English and French was sponsored by the state and involved conquest and land seizures. The numbers of migrants were constrained by contemporary technological and economic capacities, and relatively small population numbers in the colonial powers. Less than half a million Spaniards are estimated to have come to the Indies before the seventeenth century (Boyd-Bowman 1976). Colonisation gained speed after the Napoleonic Wars. Over 55 million Europeans moved to the Americas during the years 1815–1930. This period was dominated by the individual economic interests of immigrants (Baines 1998). European colonization of the Americas was followed by forced migration of Africa's population as slaves. Some 11.86 million slaves were shipped across the Atlantic in the sixteenth through nineteenth centuries and the overall death rates during the crossings were between 10% and 20% (Lovejoy 1989). European colonisation of the Western Hemisphere also wiped out over 90% of the indigenous population, due to warfare, enslavement and transmission of infectious diseases (Thornton 1987).

Chinese migration has existed over several thousands of years. It intensified in the nineteenth and twentieth centuries with the decline of Imperial China and the development of modern transport networks. There were some 31 million people of Chinese origin in 131 countries by 1990 (Poston *et al.* 1994). South East Asia used to be the traditional destination for Chinese traders and labourers until the nineteenth century but later the USA, Canada and Australia gained importance.

All these different migrations involved different risks and uncertainties for the migrants and those impacted by the movements.

Sources: Baines (1998); Boyd-Bowman (1976); Lovejoy (1989); Manning (2005); Noble and Goffart (2006); Poston *et al.* (1994); Thornton (1987); Wood (2001)

These changes in migration have gone hand in hand with transformations in the understanding or framing of migration risks. Risk is increasingly seen not so much as an object of conscious decision making, but as something that society is at risk from (Furedi 2006), a generalization that is particularly relevant to discourses about international migration (Williams and Baláž 2012). In other words, societies or communities are portrayed as being *at risk* from migration in some way, as opposed to making conscious decisions about risks relating to shortages of care workers and ageing populations, or the balance of risks between 'importing' versus training workers in order to fill skills gaps.

In reality, migration is informed by, generates and ameliorates risk, whether for migrants, non-migrants in sending communities, or populations in intermediate and destination countries. First, migration is informed or shaped by risk: it influences who migrates and who stays, the selection of destinations, the means of migration (e.g. whether intermediaries are used), the experience of the migration process, and the different stages of the migration cycle while abroad, including the decision to stay or return. Secondly, migration generates risks, real and perceived, particularly for communities and societies. This is evident in discourses about 'illegal' migration, and about smugglers and traffickers (Koser 2005; 2008), but also about individual migrants, especially those who are considered to be at risk and vulnerable, such as migrant sex workers (Agustin 2007), and live-in women care workers (Anderson 2000). Thirdly, migration can be seen as a strategy to mediate or ameliorate risk, and this is especially evident in the new economics of migration theories which conceptualise migration in terms of household risk diversification strategies (see Chapter Five). Moreover, migration is also a means of knowledge acquisition (Williams and Baláž 2008), which arguably enhanced the capacity of individuals to manage the risks associated with further migration, thereby increasing risk resilience (Alwang *et al.* 2001).

The diverse ways in which risk and uncertainty can be understood (e.g. at different scales, in different contexts), underlines the need to consider the meanings attached to the terms, and how they are theorised.

DEFINING RISK AND UNCERTAINTY

There is a wide range of definitions of risk, and Aven and Renn (2009) provide a summary of these (Table 1.1; but see also Macgill and Siu [2005]); they are mostly generic, rather than being migration-specific. The first six definitions are based on probabilities and expected values, whereas the latter are based on uncertainties. In the first four the consequences are negative, and these definitions most strongly resonate with popular understandings of risk, that is, the possibility or probability of future negative impacts or outcomes. Risk may also be associated with other terms, such as hazards or threats, which similarly imply negative impacts. In contrast, definitions 5–10 recognise that outcomes or impacts can be either positive or negative. Given the emphasis on 'impacts', it follows that many definitions understand risk in terms of the combination of the likely occurrence of an event (the probable frequency) and the severity of the outcome. For example, the likelihood of a regular labour migrant being killed in a particular destination is a severe outcome, but the probability of this is very low. Whether

Table 1.1   Ten definitions of risk: Aven and Renn

---

1. Risk = expected loss (Willis 2007)
2. Risk = expected disutility (Campbell 2005)
3. Risk = probability of an adverse outcome (Graham and Weiner 1995)
4. Risk = probability and severity of adverse effects (Lowrance 1976)
5. Risk = combination of probability of an event and its consequences (ISO 2002)
6. Risk = set of scenarios, each of which has a probability and a consequence (Kaplan 1991)
7. Risk = the two-dimensional combination of events/consequences and associated uncertainties (Aven 2007)
8. Risk refers to uncertainty of outcomes, of actions and events (Cabinet Office 2002)
9. Risk is a situation or event where something of human value (including humans themselves) is at stake and where the outcome is uncertain (Rosa 2003)
10. Risk is an uncertain consequence of an event or an activity with respect to something that humans value (IRGC 2005).

Source: Aven and Renn (2009: 1)

these frequencies and severities of outcomes are assessed 'objectively' or subjectively is a key issue that permeates our discussion in this book.

Definitions 7–10 emphasise that outcomes are not probabilities (which assumes that the risks are known and quantifiable) but are uncertain, implying various forms of gaps in our knowledge about their nature, extent and even their existence. There are many different conceptualizations of risk versus uncertainty (Camerer and Weber 1992), but the classic differentiation was provided by Knight (1921) who contended that they represent, respectively, known and unknown uncertainties.

> Uncertainty must be taken in a sense radically distinct from the familiar notion of Risk, from which it has never been properly separated. The term 'risk,' as loosely used in everyday speech and in economic discussion, really covers two things which, functionally at least, in their causal relations to the phenomena of economic organization, are categorically different. . . . The essential fact is that 'risk' means in some cases a quantity susceptible of measurement, while at other times it is something distinctly not of this character; and there are far-reaching and crucial differences in the bearings of the phenomenon depending on which of the two is really present and operating. . . . It will appear that a measurable uncertainty, or 'risk' proper, as we shall use the term, is so far different from an unmeasurable one that it is not in effect an uncertainty at all. We . . . accordingly restrict the term 'uncertainty' to cases of the non-quantitative type.
>
> (Knight 1921: 19)

In the modernist tradition, closely associated with the work of economists, risk has been understood in terms of known probabilities of outcomes (Zinn 2004b). Risky decisions, therefore, involve selecting from a range of outcomes, whose future probabilities are known (Tversky and Kahneman 1992). In the everyday world, this is best represented by gambling, where—as with roulette wheels or lottery tickets—the probable outcomes are precisely known. Zinn (2008) expresses this in terms of risk being the distinction between reality and possibility.

In migration there is always a degree of uncertainty in many key areas: will a job be found, how long will this take, what type of job will it be, and what will be the rate of pay? Even if a job has been arranged in advance, for example, by a friend or a recruitment agent, there still remain uncertainties about working conditions, or the degree to which you will feel alienated or homesick. Uncertainty in migration has two sources. The first is based on the impossibility of obtaining complete knowledge about current conditions even in the place of origin, let alone in one or more destinations, that you may never have previously visited. This can be understood in terms of Polanyi's (1966) distinction between codified and tacit knowledge. Migrants may have relatively good, although incomplete, codified knowledge about destinations, which they have obtained from reviewing web sites and books. They may also have acquired some tacit knowledge about the destinations, most obviously from having previously visited them, perhaps as tourists, or from talking to current or returned migrants. Interestingly, McKenzie *et al.* (2013) in a study of New Zealand migrants found that men significantly underestimated their potential earnings in the destination, whereas women had relatively accurate estimates. One reason for this pattern was the age-specific nature of extended family informants, who, being relatively older, were less well informed about current wages for migrants of a particular age and skill set, combined with difference in the gender wage premium in the countries of origin and destination. However, even if some migrants are able to estimate wages in the destinations relatively accurately, they lack the tacit knowledge about cultures and institutions that can only be acquired by pro-longed immersion; that is, they lack encultured and embedded knowledge of destinations (Blackler 2002; Williams and Baláž 2008: chapter three).

The second source of uncertainty is the unpredictable nature of the future. All future possible outcomes involve uncertainty, because of the impossibility of knowing the probabilities of particular outcomes in the future, even if they are currently known. Adams (1995: 29) captures this:

> Risk is constantly in motion and it moves in response to attempts to measure it. The problems of measuring risk are akin to those of physi-cal measurement in a world where everything is moving at the speed of light, where the act of measurement alters that which is being mea-sured, and where there are as many frames of reference as there are observers.

In one of the very few studies of the influence of different types of uncer-tainty on future migration decisions, O'Connell (1997) found that uncer-tainty about the future in general, that is, in both the origin and the destination, is a significantly more important influence than uncertainty about current conditions in the destination (lack of tacit knowledge) in rela-tion to the decision to relocate.

In practice, individuals act on the basis of expectations. They may seek to convert uncertainties into expected outcomes, basing these on a combi-nation of knowledge and intuition (Styhre 2004). These expectations may be expressed in terms of probabilities, and they are more likely to be in relative than in absolute terms. For example, 'I am more likely to be able to get a job in a restaurant than in an office' rather than 'I have a 40% prob-ability of getting a job in an office'. Or they may explicitly recognise uncer-tainty: 'I have no idea what type of job I will get, but I will just take a chance'. Or they may rely on intuition: 'I don't know, but I feel that I will be able to get a job in an office'. These uncertainties are compounded across the different domains of the migration experience, including not only work but also family, community and culture (Williams *et al.* 2011).

Migration is associated with uncertainty because of the highly imperfect, highly uneven, and contested nature of the knowledge migrants possess about outcomes in possible destinations. That is not to say migration occurs under conditions of total uncertainty. Migration decisions are informed by *some* knowledge about the destination. As the authors have written elsewhere (Williams and Baláž 2012: 3):

> Migration should probably be understood as being associated with expectations about risk formed under conditions of partial knowledge. This is best understood as a continuum of knowledge and uncertainty, or of risk and uncertainty, which is fluid with individuals moving in both directions along the continuum in terms of personal understandings of the limitations of knowledge.

These definitions are, of course, essentially rationalist, in so far as they assume that individuals know there are risks and uncertainties involved in decision making. In other words, they know that there are limits to their knowledge, even about not knowing. But do migrants know that it is impossible to know—that they are making decisions without knowing about, or even suspecting the existence of, major influential factors? Or do they make their decisions, assuming they know most of the key factors, whether or not they have sufficient knowledge to formulate even approximate probabilities of the likelihood of different outcomes? It is a question with profound implications for the way that migration decision making, and behaviour or practices, are conceptualised and analysed.

While the central focus of this volume is on risk and uncertainty, we also need to consider another important concept: trust (Siegrist and Cvetkovich 2000). Trust plays a central role in mediating risk, whether in terms of the responses of host societies or individual migrants and their families (Zinn 2004a; Nuissl 2002). As Lewis and Weigert (1985: 463) powerfully assert, 'Trust begins where knowledge ends'. For Macgill and Siu (2005: 1108) trust is 'a means of coping with limited knowledge or complete lack of knowledge (incertitude), and also . . . a predominant determinant of risk acceptability'. Given complete knowledge, arguably there is no need for trust, as the rational individual can make decisions based on all 'the facts', assuming that risk is knowable and measurable. In practice, all actions are based on partial knowledge and therefore depend to some degree on trust (Tierney 1999). The extent to which the notion of trust is consciously or unconsciously articulated, and based on intellect (rational assessments) as opposed to intuition (Styhre 2004), is variable. But one thing is evident—trust is a particularly important issue for migrants in the face of risk and uncertainty, and having limited tacit knowledge about migration destinations.

This takes us to the question of how trust should be conceptualised and defined. There is a fundamental difference in the approach of economists and sociologists. Economists reduce trust to transaction costs: the transaction costs of migration are increased in the face of a lack of trust in the knowledge available about likely migration experiences and outcomes. Rationally, economists would expect higher returns from migration decisions based on trust to compensate for the increased transaction costs. Behavioural economists, however, consider this to be oversimplified, and have identified source preference as a heuristic used by individuals when making decisions in the face of incomplete or competing knowledge (Fox and Tversky 1995). Faced with a choice between two destinations, a migrant may decide to migrate to the country with lower wage rates, contrary to the expectations of classical economics, because the source of information about potential wages in this country is a trusted friend. Such models, however, remain essentially rationalist, even if the basis of decision making is more nuanced than in classical economics.

In contrast, sociologists mostly consider that trust is something that is given in advance (of, say, the migration decision-making process) and is rooted in shared values and routines (Anheier and Kendall 2002: 347). That trust may arise from collective allocation of trust in particular individuals or sources (perhaps a narrative about a 'famous' member of a community who has achieved success or fame through migration) or through interpersonal relations—trusting the advice of particular individuals, as a result of sustained social interaction over a period, probably across a number of domains. The media also increasingly play a key role as sources of knowledge about particular events, although such trust is variable (Lupton 2006; Tierney 1999). Information is critically evaluated in terms of whether the sources are considered to be biased in some way, or to be unduly informed by vested interests (Slovic *et al.* 1979). Hence, there is particular value attached to information provided by trusted friends.

RISK AND MIGRATION: MULTI-SCALAR PERSPECTIVES

Some of the ambiguities relating to the role of risk in migration stem from its scalar, or multi-scalar, nature (Williams 2007). Brettell and Hollifield (2008: 9) argue that:

> Objects of inquiry and theory building are closely related to the levels and units of analysis. In migration research, these vary both within and between disciplines. An initial contrast is between those who approach the problem at a macrolevel, examining the structural conditions (largely political, legal, and economic) that shape migration flows, and those who engage in microlevel research, examining how these larger forces shape the decisions and actions of individuals and families, or how they effect changes in communities.

Risk can also be thought of at a number of scales which are considered here in terms of the necessarily inter-related individual, household, community, regional, national and global levels. These levels will be discussed separately in order to highlight key issues, and different foci at each scale, but they are necessarily inter-related because, as Amin (2002: 386) argues, space, place and time are co-constituted and folded together. For example, the national level conditions the rights to work and residence which shapes the risks for individual migrants in terms of employment and social integration. But it is also true that events (such as attacks on migrants, or social unrest articulated in anti-migrant terms) influence national regulatory changes, designed to mediate perceived risks associated with migration.

Individuals

For individuals, the fundamental question is how to balance the risks of staying versus migrating. Migration is popularly presented as a risk-taking activity and it is true that it is risky, or uncertain, given that potential migrants have less knowledge of the destination than they have of their current place of residence. However, there are risks associated with both staying and migrating. In extremis, it is not necessarily the most risk-tolerant individuals who migrate. Sometimes economic or political conditions in the home area may be so challenging that staying represents a higher level of expected negative risks. Hayenhjelm (2006) considered this generically in relation to why, say, refugees pay smugglers for dangerous boat journeys across borders, or why individuals will take the health risks of selling a kidney. The answer lies in what are termed 'risks from vulnerability':

> If an individual takes a large risk because of an underprivileged position and due to a lack of positive alternatives, or is led to believe that the risks are smaller, or the benefits greater, than they in fact are, because that person is not able to verify or falsify that information, then this is a case of risk from vulnerability.
>
> (Hayenhjelm 2006: 190)

Vulnerability is not the same as victimhood. Indeed, this is an approach that recognises both structural constraints (lack of resources, including knowledge) and the role of human agency. Even in extremely difficult conditions, individuals decide (which is not the same as choice) whether to migrate or not to migrate, and there are risks associated with both pathways. However, when risks from vulnerability exist, staying may actually pose a greater risk than migrating.

The risks associated with migration are, of course, dependent on the nature of the migration. They are relatively low for, say, the employee being transferred to a different international branch by a multinational company, or for a student going abroad on an ERASMUS year abroad. In contrast, they are significantly greater for the labour migrant travelling abroad in the hope of, rather than a guaranteed, job. And they are far greater for the individual trusting his/her life or future to a smuggler who clandestinely crosses the border, often at high risk of death or detection, or for the refugee fleeing potential starvation or torture. One of the advantages of focusing on the risk involved in these different forms of migration is that it highlights that all forms of migration involve risks to varying degrees. It also recognises that human agency is exercised in all these different forms of migration: even when individuals are most vulnerable, there are still decisions to be taken about whether to migrate, which route to follow, which smuggler to trust, or when to leave. This is not to say that the diverse strands of migration research can be unified around conceptualizations of risk, but it does provide insights into some of the more problematic questions, such as deciding when migration is voluntary or involuntary, and it recognises a continuum of risks in all forms of mobility, rather than simplified dichotomies.

The risks associated with migration vary across the migration cycle. They will be different in the pre-migration stage (who is at risk, what are the risks of non-migration), during the migration process (risks from travel accidents and crime, or risks to health), in the initial period at the destination (of finding jobs, of being homesick, of getting decent housing) and in the longer term (risks of social exclusion, of entrapment in the bottom end of the labour market, of children growing up with different cultural values and not speaking their parents' first language). None of these risks are predictable, but involve a mixture of risk and uncertainty.

Risks also differ between individual family members over time, as can be illustrated with respect to health. The most obvious risks in the migration cycle are encountered during the process of migration itself, when migrants may be vulnerable not only to diseases, but to death or injury from dangerous crossings, or from assault. For example, Médecins Sans Frontières (MSF 2005) reported that a quarter of the medical consultations they provided to irregular migrants in Morocco involved acts of physical violence against their persons. The risks do not end after reaching the destinations, as migrants are subject to higher levels of risk than the indigenous population, on average, both in the domains of work and everyday life (Voluntary Health Association of India 2000). They are often disproportionately concentrated in the most hazardous occupations, and to live in unsanitary conditions. Migrants are also considered to be unduly at risk from AIDS and sexually transmitted diseases due to being away for their families for prolonged time periods. Eventually migrants may obtain better paid jobs, and be able to improve their housing. However, they still tend to receive less health care than non-migrants (Derose *et al.* 2007), because of language barriers, lack of knowledge of their rights, and—if undocumented—access constraints. The particular risks associated with mental health are possibly even more complex (see Box 1.2)

Box 1.2   Migration and Mental Health Risks (Schizophrenia)

Migrants seem to be more at risk than the indigenous populations in the destinations from schizophrenia. The higher recorded rates may be due to several reasons, including differences between sending and receiving countries, the selectiveness of migration, the migration process itself, and cultural differences in how schizophrenia is reported and diagnosed.

1. Sending countries have high rates of schizophrenia.
2. Schizophrenia predisposes people to migrate.
3. Migration produces stress and elevated rates of schizophrenia.
4. High rates can be explained by misdiagnosis.
5. Symptom differences.
6. Ethnic density may play a role in high rates.
7. Concepts of self may influence high rates.
8. Aspirations and achievement disparity may influence high rates.

Source: Bhugra (2004)

Households

Migration usually has substantial risk implications not only for individuals but also for their family and household members. Although many migrants are young and single, many have partners, children and an extended family. The key decision-maker may be one adult, while other adults and children are to varying degrees passive or active co-decision-makers. Children may exert some, or even considerable, influence on the decision-making process, but are rarely actively incorporated into this. All these different individuals—lead migrants, following migrants, and children—face different forms of risk. The risks faced by a non-working partner (usually a woman) in terms of providing child care, or working part time, may be different to those encountered by the lead migrant, expected to provide the main source of family income, but they are no less substantial. The same applies to the risks faced by children, starkly manifested by their first day at school, perhaps in a very alien culture or educational system, who are expected to understand a different language while lacking their previous supportive network of friends.

'Trailing migrants' are a diverse group, and include not only partners and children but also grandparents, other relatives and, exceptionally, care and domestic service workers. They, and the risks they face, have been relatively neglected in the migration literature, as part of the more general neglect of family migration (Kofman 2004). However, gender differences have attracted greater research attention in more recent years, and there has been a debate about the career risks for 'trailing spouses' following 'breadwinner' lead migrants. Clark and Withers (2002) consider these to be significant, not least because of visa restrictions imposed on accompanying family migrants, as well as the transferability of their skills and knowledge. However, trailing spouses are not necessarily passive victims, as they may be willing to take these risks, and aspire to developing new career trajectories. However, it is difficult not to agree with Bruegel's (1996: 250) conclusion that 'at most the breadwinner model may have been modified rather than transcended' in terms of the distribution of risks.

There has been little formal conceptualization or research on the distribution of risks across families and households with one striking exception, the so-called new economics of migration (Stark and Bloom 1985), which addresses household risk diversification strategies. These are considered further later in the book, and it is sufficient here to note that, contrary to neo-classical economics, they contend that households are not only concerned to maximise their total incomes but also to reduce risks to those incomes (Stark and Katz 1986). The theory was developed against the background of volatile economic conditions faced by households working in agriculture in less developed economies, where livelihoods are routinely uninsured, by either the state or the private sector, against natural and other disasters. Migration to an 'uncorrelated' labour market, whether in the same country or abroad, becomes a means of diversifying, and reducing the risks, faced by the household. However, such migration is not risk free. Rogaly and Rafique (2003) found that the family members who stayed became dependent in two ways. First, on the remittances of the (usually) male migrants, which are subject to employment and health risks in the destination. Secondly, because of delays in the receipts of remittances, they become dependent on borrowing food and asking for other forms of assistance from other relatives in the village.

The new economics of labour undoubtedly brings new perspectives to migration studies, but it is also problematic, tending to conceptualise the household as a single, uniform decision-making unit. In reality, households are constituted of diverse interests and age- and gender-specific power inequalities. Individual decision making is equated with serving the overall household interests, whereas in reality it may be driven by particular individual interests. In other words, the household is seen as 'an individual by another name' (Folbre 1986: 5). As Wolf (1990: 43–44) argues:

> Feminists have cut through romantic assumptions about family and household unity, arguing that there exist instead multiple voices, gendered interests, and an unequal distribution of resources within families and households. . . . Attention is slowly turning to intra-household relations between genders and generations . . . yet we still know relatively little about intra-household processes, conflicts and dynamics, particularly within poor Third World peasant and proletarian households. Indeed, Third World household studies appear to be the only context in

which the myth of family solidarity and unity is perpetuated, and this is seen most clearly in the concept of household strategies.

The myths of solidarity and unity are particularly striking when considered against the distribution of risks across the household.

Communities

There is a considerable gap between individual and broader community risks relating to the costs and benefits of migration. The aggregate effects of individual migration on the economy of the community of origin can be substantial. There are implications in terms of the supply of labour, skills, and wages, although these are all highly contingent on the nature and scale of migration, and the prevailing labour market conditions in that community: is there surplus labour, potential to substitute capital for labour, or to source labour via in-migration? Over and above the aggregate economic effects, there are three main impacts to consider at the collective level: the maintenance of facilities and services, demonstration/networking effects on migration intentions in communities of origin, and social inclusion in destination communities.

There are important risks for the sustainability of services and facilities in the sending communities. There is a positive side, with evidence that migrant hometown associations abroad can be a source of collective funding for communal facilities in villages, such as schools and health clinics, and provide medical supplies or improved water supplies. They also provide support for particular religious festivals, or to repair local historic monuments such as churches (Orozco 2005). There is no guarantee of such outcomes—instead, there is a risk attached to the positive outcome. The probability of that risk is measurable. For example, in the USA the percentage of remittance senders who also belong to hometown associations varies from 1% for migrants from Bolivia compared to 29% from Guyana (Orozco and Rouse 2007).

In contrast to these positive outcomes, there is the negative risk that outmigration may lead to highly age-unbalanced and declining populations which are unable to support local services, whether schools, health care, or agricultural systems which require large inputs of communal labour (e.g. for maintaining irrigation or terracing systems). These risks are borne by all communities with high levels of outmigration, and apply to more developed as well as less developed economies, as illustrated by southwest Alaska (Box 1.3).

Box 1.3   Outmigration, Population Decline and School Closures in Southwest Alaska

Migration has contributed to a general population decline throughout large parts of southwest Alaska over the last decade. The extent of the decline was greatest in Bristol Bay, where it amounted to –23% between

2000 and 2009. Moreover, 'while the overall impact on the broader base of communities is stark, population declines have even greater ramifications for small villages where large outmigration has occurred, often cutting community populations in half in less than a generation's time'. For example, the population of Chignik Lake declined by 50 from 145 to 73 in less than a decade.

The greatest risk that this poses for individual communities is that school policy dictates that schools require a minimum of 10 pupils if they are to remain open. As school enrollments have dropped by some 55% between 1997 and 2010 in Bristol Bay Borough, this has led to school closures or to schools fast approaching the minimum threshold levels with the risk of closure in the near future. Closures of schools have broader ramifications, as they can trigger further outmigration, and this in turn can pose risks for other village services such as regular mail deliveries, or maintenance of air links.

Source: Pebble Partnership (2012)

Another important set of risks are related to the cumulative nature of migration. This works in two ways—via a demonstration effect, and through creation of networks which facilitate future migration by reducing the transaction costs and risks of negative outcomes. There are also two main negative risks attendant on this. First, that it may facilitate high levels of outmigration that undermine the provision of local services and facilities, as noted above. Secondly, that it creates a culture of dependence on migration as a solution to economic development, distracting attention from indigenous development initiatives.

There are also risks for migrants and communities in the destinations. These include economic risks that are considered in the following discussion of the national scale, but they also include issues around social inclusion and social cohesion. Much of the debate is couched in terms of assimilation or the acceptance of the culture of the destination community. However, Hickman *et al.* (2012) challenge this, arguing that social cohesion is the outcome of migrants being able to resolve day to day conflicts (and risks) in positive ways. There is no inevitability in the outcomes, but rather there are deep uncertainties as to how the contours of local economic development, histories of migration and narratives about these, and individual experiences of risk and uncertainty are intermeshed.

National

Much of the debate about the risks of migration at the national level has focused on economic issues, although there are also popular discourses about the impacts of migration on national cultures, which are considered in Chapter Nine. The economic analyses of the impacts of migration are complex and highly contingent, involving not only individual wages, savings and remittances, but also collective or public expenditures and taxation receipts, as well as issues around inflation, productivity and multiplier effects. The economic impacts of migration tend to be highly uneven and uncertain, even at the national level. This leads Taylor (2006) to conclude that 'researchers used to ask whether migration has a positive or negative effect on development. Today they are more likely to ask: "Why does international migration seem to promote economic development in some cases and not in others?"' Similarly, there are risks attached to policies that promote or seek to limit international migration, as evident in the UK's Migration Advisory Committee's periodic reviews of skills shortages (Box 1.4)

Box 1.4   Evidence of Risks of Shortages of Skilled Workers in the UK Resulting From the Capping of Migration Numbers

'We typically have anything between 50–100 vacant nursing positions at any one time. This shortage of nurses at times means we are unable to accept the number of patients requested by the referring NHS hospitals, thus creating additional pressures on local NHS Trusts, with *the risk* of patients being unable to access life-saving treatment.'

Fresenius Medical Care response to MAC call for evidence

'The oil and gas industry is currently in a situation where failure to obtain the right skills could undermine the North East's plans to promote itself as a global hub of expertise and *risk* damaging the Scottish and UK energy sector at an important time for the economy.'

Scottish Council for Development and Industry response to MAC call for evidence

'The reality is that Rolls-Royce will find it extremely difficult to recruit engineers in the UK market in the next few years and will need to recruit from overseas. If Rolls-Royce is not able to recruit migrant workers, there is a potential *risk* that they will need to increase their overseas production capacity to meet demand.'

ADS response to MAC call for evidence

Source: Migration Advisory Committee (2011), italicization added.

There has been a strong academic focus on skilled labour migration, and, although rarely explicitly couched in such terms, risk and uncertainty are implicit and pervasive in this literature. Table 1.2 identifies six main possible outcomes in human capital terms. It also identifies differences in the risks associated with each of these, which are expressed at different scales for individual migrants and home versus destination countries.

Table 1.2   Risks associated with different types of human capital redistribution via international migration

'Brain' Redistribution	Human Capital Outcomes	Associated Risks
Brain exchange	Relatively balanced (mostly temporary) flows between core economies whereby, implicitly, effective use is made of human capital.	Risks, both positive and negative, are relatively balanced in line with the migration flows.
Brain drain	A (permanent) transfer of human capital from less- to more-developed economies.	Risk to the countries of origin.
Brain overflow	The (permanent) transfer of human capital that is underutilized in countries of origin.	Risk for the migrants, as the national economies of origin are currently not fully utilizing their skills.
Brain waste	Ineffective utilization in the destination (or origin) of human capital (permanently) transferred from the origin.	Risk to the migrants themselves, and to the countries of destination and origin
Brain training	Human capital enhancement via mobility specifically for educational or training purposes.	Risks are potentially positive for the migrants, although there are risks of failure. There are positive or negative risks for the origin and destination countries, depending on whether and when these migrants return to the former
Brain circulation	Human capital enhancement via (temporary) mobility which, implicitly, is used more effectively upon return	Risks of temporary transfers of skilled workers for the sending countries, but with indeterminate conclusions as to whether the outcomes are positive or negative for the sending/receiving economies or the individual migrants.

Source: Lowell and Findlay (2002); Ghosh (1985); Williams and Baláž (2005)

Finally, we re-emphasise that although the risks attached to migration are articulated in different ways at different levels, these are enfolded. The risks to the individual migrant overlap with those for the household: economic success for, and significant remittances from, the migrant provide a form of insurance for the household that remains in the sending country, but the households are also subject to other risks, such as dependency and an uneven distribution of the costs and benefits of migrants, which are often highly age-specific and gendered. The outcomes for communities and national states are in part determined by the aggregate economic outcomes for individual migrants and households, but they are more than this because of the impacts on the collective domains, whether in terms of hometown associations' communal contributions, social cohesion, health and educational consequences, public expenditure, labour market displacement, total demand, inflation or taxation. Moreover, the openness of communities and national economies to migrants, and the nature of national regulatory regimes, significantly shape the opportunities and risks faced by migrants.

THE APPROACH OF THIS VOLUME: CONTRASTING APPROACHES TO THEORISING RISK IN MIGRATION RESEARCH

While migration research often comments on the existence of risk (Massey *et al.* 1993; Roberts and Morris 2003), there is little *explicit* theorization of risk/uncertainty in migration. At worst, it is ignored, and more commonly it is implicit, for example, in discussions of health issues or wages. There is also considerable confusion in the use of the terms risk and uncertainty, and the way these are applied at different scales. This largely reflects the lack of, and uneven, theorization of migration risks, as well as the different theoretical perspectives brought to the subject by, say, economists compared to geographers or sociologists. As in a different field of migration research, internal versus international migration, 'dichotomisation seems to have been influenced by several factors, including different data sources, different disciplinary backgrounds of researchers, different analytical techniques, and different research agendas that reflect different policy concerns and funding sources' (King and Skeldon 2010: 1620).

There are fundamental divides between the rationalist and social constructionist approaches to risk, and more generally between the approaches of economists (and psychologists) on the one hand, and of sociologists and anthropologists on the other. Our aim in this volume is not to advance risk as a unifying concept for the field of migration studies, but to sketch out the prevailing theoretical perspectives that have been used, or have potential to be applied, at different scales. We are interested as much in the linkages between these, as we are in the difficulties of reaching across scalar and theoretical divides that represent deep chasms in our understanding of migration, risk and uncertainty.

The remainder of this section explores the value for migration studies of rationalist and social constructionist approaches to theorising risk and uncertainty, drawing especially on economics and sociology, but setting them in the context of different scales. The former assume that risks are objective and 'out there', and that they can be quantified; they also 'recognise the existence of uncertainties, which cannot be quantified. The following discussion draws heavily on our previously published theoretical overview (Williams and Baláž 2012), and that in turn had been significantly informed by the work of the ESRC's SCARR (Social Contexts and Responses to Risk) programme on risk research (http://www.kent.ac.uk/scarr/). Somewhat surprisingly, the SCARR network paid relatively little attention to migration.

Individual Scale: Rationalist Theoretical Approaches, Cultural Theories, Intermediaries and Networks

Migration and risk has probably been subject to more explicit attention and theorization in economics than in any other discipline. Consistent with the view of risk being 'out there', measurable, and subject to rational decision making, most economic theories focus on risk rather than uncertainty. In practice, there are considerable challenges in measuring risk, and economic modeling often relies on surrogate measures. Decision making is assumed to be rational and to have clear goals, whether these are the maximization or satisficing of returns versus costs, or risk diversification.

In Chapter Two we review the contribution of *classical and neoclassical economic theories* to the analysis of risk. There is a well developed literature on migration within this theoretical framework (see Todaro 1969; Todaro and Maruszko 1987), but it pays only limited attention to risk. Instead, neoclassical approached focus on utility, as the net outcome of the costs and benefits of migration to individual migrants. Conceptually, such theories do acknowledge the role of risk, understanding outcomes as probabilities rather than as given. Massey *et al.* (1994: 701, emphasis added) summarise this approach:

> Expected income is defined as the *probability* of employment (one minus the unemployment rate) times the mean income in whatever economic sector a rational actor contemplates working. For undocumented migrants, this product also needs to be multiplied by the *probability* of successfully entering the destination country and evading deportation (one minus the *probability* of getting caught). The difference between incomes expected at origin and destination, when summed and discounted over some time horizon and added to the negative costs of movement, yields the *expected* net gain from movement, which if positive promotes migration.

Human capital theories represent an important extension to this approach, and have been extensively applied to migration studies. The individual migration decision is conceptualised as an investment decision based on the returns to their human capital in different regions or countries. These potential lifetime economic returns are balanced against the *known and unknown* costs of migration (Stark 1991), encapsulating both monetary and non-monetary costs and returns. At the conceptual level, both risks (known) and uncertainties (unknown) are acknowledged but they are not explicitly analysed; rather they are assumed to be factored into estimated future incomes and costs (Katz and Stark 1986). Even when risk is recognised as being potentially important, it is assumed away with individuals being considered 'risk neutral' (Beine *et al.* 2001: 279). The final part of the chapter examines how economic theorists have tried to address the limitations of human capital and neoclassical theories. Tversky and Kahneman's (1992) prospect theory has potential for migration studies, especially because of the importance of risk aversion and potential loss as a deterrent to migration, given the substantial gains and losses associated with mobility.

In Chapter Three, we discuss the attempts of *behavioural economists* and psychologists to move beyond the limiting assumptions of the neoclassical approach, by engaging with *bounded rationality*. The work of the Nobel Prize-winning economists Kahneman and Tversky has been particularly important in providing insights into how individuals make decisions in the face of uncertainty and incomplete knowledge (Tversky and Kahneman 1974). A number of heuristics developed in generic research in behavioural economics can usefully be applied in migration studies. For example, 'anchoring' (Kahneman and Thaler 2006: 223) could be applied to making a decision about migrating to a destination in the face of incomplete knowledge about it. This could involve initially anchoring the assessment onto the data that is available, say for a different location which is assumed to have similar features.

Behavioural economics and economic psychology also emphasise the importance of individuals' psychological profiles (Mehta 2007), and of particular interest are the notions of risk aversion and risk tolerance (Tversky and Kahneman 1974). Massey *et al.* (1993: 456) argued in their seminal review of migration theories that it has remained difficult to express the probability of migration convincingly 'as a function of individual and household variables'. In contrast, notions of risk aversion and risk tolerance provide insights—however partial—into why some individuals within these groups are more likely than others to become migrants or to stay (for example, Jaeger *et al.* 2007; Dohmen *et al.* 2005; Heitmueller 2005). They also provide insights into whether migrants are more or less able than non-migrants to tolerate uncertainty as opposed to risk (Baláž and Williams 2011), and the importance of competence to manage risks (Williams and Baláž 2013).

Chapter Four explores further how individuals make decisions when faced with either incomplete or vast amounts of knowledge. At the heart of this is the conceptualization of *complex decision making*. This approach remains grounded in the notion of bounded rationality, and therefore shares some of the reductionist features of the rationalist theories discussed earlier. This approach has particular relevance for migration studies, where individuals are faced with potentially vast amounts of information about a very large numbers of potential destinations, but at the same time there are significant gaps in the information that is available in codified form, let alone in terms of tacit knowledge. Drawing on Baláž *et al.* (2014), we consider how individuals select and weight a range of economic and noneconomic factors, are influenced by the image attached to the names of particular countries, and make decisions in the face of uncertainty. This represents one of the few applications of experimental research methods to understanding migrant decision making in any circumstances, let alone in relation to risk and uncertainty.

In Chapter Five the focus shifts from the individual to the household. Arguably, the work of the 'new economics of migration' on *household risk diversification strategies* represents the single most coherent, explicit and well developed strand of conceptualization of risk in migration studies (Stark and Bloom 1985; Katz and Stark 1986). The key difference to neoclassical theories is that the decision-making unit is assumed to be the household, and this means they consider the spatial and sectoral distribution of household labour in terms not only of maximizing economic returns but also in reducing the risk and uncertainty that household income is exposed to. This is particularly important in economies, such as the rural sector in less developed economies, where there is either weak or no system of social insurance, and individuals are unable to access private insurance schemes. In these circumstances, the risk to household incomes (from natural disasters, or sharp declines in market prices) can be calamitous. Households respond to this by seeking to distribute their total income over a range of sources, some of which may involve migration. The key point is that these income sources should be 'un-correlated'—leastways, as far as it is possible, in an increasingly interdependent global economy. However, there are also several limiting assumptions about the household being a unified decision-making unit, understating the risks of being dependent on individuals whose loyalty (and remittances) to the household may change over time, and increased risks to local communities. In common with all the economic approaches considered up to this point, it understands risk as real, measurable, and 'out there'.

Over time, economists have developed a number of alternative theoretical approaches that seek to move away from the assumptions of neoclassical theories and to engage with the notion of bounded rationality. Sociological theories mostly adopt social constructionist approaches. Risk is understood to be 'discursively constructed in everyday life with reference to the mass media, individual experience and biography, local memory, moral convictions, and personal judgements' (Zinn and Taylor-Gooby 2006b: 54). They also mostly shift the focus of the analysis from the individual/household to the collective.

Chapter Six considers how risk is understood in *culturally based theories*. The essence of this approach is that 'the individual's perception and response to risk can only be understood in context of their embeddedness in a sociocultural background and identity as a member of a social group, rather than through individual cognition' (Zinn and Taylor-Gooby 2006a: 37). Risk is considered to be socially constructed and socially situated. Moreover, there is a focus on relationships with other individuals or communities, and an emphasis on the need to take a longitudinal or life course perspective, recognizing that individuals accumulate learning about risk and personal risk management over time, and through highly time- and place-specific social interactions (Macgill and Siu 2005). Similarly, Boholm (2003a) considered that 'in real life situations, the boundary between certitude and uncertainty is of course seldom razor-sharp, and vagueness and ambiguity tend to be the rule rather than the exception' (Boholm 2003a: 168). Probably the most significant contribution to this approach is the research of Douglas, and her collaborator Wildavsky (Douglas and Wildavsky 1982: 6–7), who challenged what previously had been a dominantly objective approach to risk (Zinn 2004a): for them it was 'a social construction in a particular historical and cultural context' which was open to varying social interpretations. These authors did not specifically address *migration* risks, but their conceptualizations have application in this field. Some commentators, however, considered that Douglas's work is too schematic in the face of the complexity of how individuals and social groups understand risk (Zinn and Taylor-Gooby 2006a). Recent research has focussed more on identities and membership of subcultures (Tulloch and Lupton 2003), and this has greater resonance with research on the role of identities in migration research, if not specifically with migration risks.

In Chapter Seven, the analysis shifts to *how individuals mediate risk through the use of intermediaries*. There are a number of different theoretical perspectives on this issue, ranging from the rationalist approach of the behavioural economists (for example, Tversky and Fox's notion of 'source preference') to cultural theories of how reliance on intermediaries is influenced by the construction of shifting social identities. Intermediaries have two important dimensions: their potential to reduce the transaction (total) costs of migration, and the extent to which individuals trust them. In migration, as in many other areas of life, trust is critical when faced with a lack of knowledge and, therefore, risk and uncertainty (Lewis and Weigert 1985). There are, of course, important differences in how trust is theorized as noted earlier in the discussion of definitions.

The most obvious form of reliance on intermediaries is via informal networks of families and friends (Haug 2008), who provide information, housing support and assistance to find jobs. There is interesting research in migration studies as to whom individuals are more likely to trust, and whether they are family or friends, although the internet is transforming trust relationships. There are also considerable differences in the way risks are mediated according to whether migrants are acting 'independently' in the destination or are drawn into established migrant communities. Such enclaves can mediate many of the risks associated with migration, such as finding a job, accommodation, or familiar cultural practices. However, enclaves are also potentially sources of other risks, particularly of exploitation in ethnic businesses, and of relatively closed social networks; in terms of occupational and social mobility, they can be both stepping stones and dead ends. Alternatively, migrants can also rely on employment and migration agencies to find jobs. Such agencies may mediate but they do not, of course, dispel the risks and uncertainties faced by migrants. Insurance also plays a role in mediating risk.

In contrast to relying on legalised and regulated agencies, migrants may rely on agencies which are operating in the grey areas of regulation, or completely outside the law. These attract far more media and policy attention. They are epitomised by smugglers who offer, at a price, to reduce or transform the nature of the risks associated with illegal border crossings (Koser 2008). Smuggled migrants are often portrayed as being highly vulnerable and at risk, and sometimes as victims, and they can be. However, as the work of Agustin (2007), Mai (2008), and Anderson (2008), amongst others, indicates, reality is complex, and the degree of self-determination and agency is variable even in such precarious occupations as migrant sex workers. Finally, the trafficking of people is a practice which inherently creates extreme risks and uncertainties for migrants, rather than mediating it.

MACRO OR SOCIETAL APPROACHES TO MIGRATION AND RISK: DOMINANTLY SOCIALLY CONSTRUCTED

Chapter Eight considers what is probably the best known of all theories of risk, Ulrich Beck's *risk society* thesis. Strictly speaking, it sees risks as both real and socially constructed. While Beck recognises that major risks existed in earlier periods, for example, from plague and famine, risk has become an 'expression of highly developed productive forces. That means that the sources of danger are no longer ignorance but knowledge; not a deficient but a perfected mastery over nature' (Beck 1992: 183). The rapid expansion of new technology has outgrown the capacity of society to manage the associated risks. Moreover, risks have become less territorially and socially bounded, and more globalised. Beck does not address directly migration issues, but we can note Tulloch and Lupton's (2003: 41) critique that it is overly focused on the 'cataclysmic democracy' of catastrophic environmental hazards, and unable to engage with how societies deal with, amongst other things, 'the risk associated with mass immigration'. While we concur with Tulloch and Lupton's comment about the emphasis on cataclysmic events, the thesis may have some leverage in relation to climate change and environmental refugee movements (Piguet, 2008). However, the emphasis on cataclysmic democracy, whereby all social groups are affected and are not protected by income, class or power, is less convincing. These, and other social characteristics, are integral to how migration is used as a response to such threats to life and livelihood.

Beck's individualization thesis appears more promising for migration research. Essentially, this argues that the reduced importance of class and estate-specific knowledge has led to greater individualization. Individuals have become more reflexive and have taken more responsibility for producing their own biographies and CVs. This resonates with Rose's work (1996a; 1996b) on being resourceful as part of the 'motives of self fulfillment', and Lyng's notion (2008) of 'edgework', whereby individuals positively value risk. In part this is due to the 'seductive power of the risk experience' (120) but it also provides opportunities to develop personal competences in managing risk and uncertainty. This resonates with many forms of student migration, gap year, and self-discovery types of migration, whereby individuals aspire to enhance their self-esteem, and their status in the eyes of their peers, and perhaps employers (King and Ruiz-Gelices 2003; Baláž and Williams 2004a).

In Chapter Nine, attention shifts to *policy and popular discourses about migration* and risk. Douglas (1992: 7) wrote that 'certain marginalised groups are identified as posing risks to the mainstream community, acting as the repository for fears not simply about risk but about the breakdown of social order and the need to maintain social boundaries and divisions'. These risk concerns feed into the policy domain, and Jennings (2007) and others demonstrate how notions of risk have driven national immigration controls and asylum policies in particular countries. These risks are amplified in power-informed discourses and they become understood to be systemic risks in complex societies, as in the case of migration in modern societies. One approach to understanding these discourses, which has particular leverage in migration studies, is Foucault's notion of governmentality, that is, on the practices rather than the institutions of governments. Foucault (1991) argues that power is not so much concentrated in governments as widely distributed across society through the practices and discourses that produce knowledge. In general, a specific event does not constitute 'a risk but its description as part of a risk calculation make it a risk' (Zinn 2007). Moreover, 'migrants at risk' or 'communities at risk from migration' are generalised social categories which are produced in media and other popular discourses, and which portray misleadingly homogenous groups while neglecting the diversity of risks faced by individuals and communities. In migration studies, these ideas apply particularly to the categorization of groups such as refugees, illegal migrants and especially to trafficked individuals.

Finally, Chapter Ten considers how modern states have sought to manage risk, and how this was rooted in modernism and notions of the scientific capacity for risk management. Initially this was expressed through the belief that modern states could calculate the probabilities, or risks, of changes in, for example, birth rates, the extent of epidemics, or unemployment across the business cycle. It is a perspective, or approach, which has come under profound assault in recent years, from both persistent evidence of the limitations of the state to predict, prevent and react to risks, and from the rise of neoliberalism. Deleuze (1995) provides one useful perspective on how uncertainties are translated into manageable risks in modern societies.

The chapter examines the extent to which understandings of risk, and of the ability of the state to manage risk, have shaped migration policies over the years, including the recent focus on 'managed migration' (Ghosh 2000). It draws especially on the work in international relations, risk and security, and on the analysis of policy making. Adamson's (2006) work is especially useful in showing how, since the terrorist attacks on September 11, questions relating to migration and security have come to be seen increasingly as issues relating to international terrorism. New technologies play an important role in border controls and surveillance of 'suspect' migrants. Ultimately this raises question about the extent to which the monopolization of control over the legitimate means of international migration by the state has itself become a major source of risk for migrants.

2 From the (Neo)Classical Approach to Prospect Theory

INTRODUCTION: BASIC CONCEPTS OF VALUE AND UTILITY

This chapter starts with the initial building block for much of the research on individual migration, the neoclassical assumption of rational decision making and perfect information. The assumption of perfect information necessarily contradicts the notion of uncertainty, although it does accept that there are risks attached to expected outcomes. However, it approaches these through assuming that they are essentially reflected in individual estimates of the costs of and returns to migration, rather than problematising risk and uncertainty. Later theoretical developments recognised differences in risk aversion.

The key issues here relate to understandings of utility. According to Bentham (1789) and the marginalist school, individuals can judge the nature and dimension of experienced utility of a particular outcome, and by extension this can be applied to migration to a particular country rather than another. As such, there is little room for explicit engagement with the nature of risk and uncertainty in such a formulation. In contrast, the expected-utility approach typically reduces outcomes to a function whereby expected utility is calculated as the product of probability and the value of the outcome, adjusted by the risk aversion of the individual. In other words, the migrant weights the expected wages or other returns in a destination by the probability of being able to realise this, and his/her risk preferences. This is an objective rational calculation, but it does acknowledge risk.

Bernoulli (1986), followed by von Neumann and Morgenstern (1944), turned value to utility ('moral expectation' in his words, Bernoulli 1986: 24), which is a more subjective notion. For example, it is not always constant as is evident in concepts such as diminishing marginal utility. In terms of migration this means the utility a migrant obtains from migrating to a particular county may not grow proportionally. The initial experiences or higher income may provide considerable utility, but this may not grow proportionally over time. Individuals, are still considered to be rational decision makers who are able to know and to estimate accurately the probabilities of outcomes.

Whereas expected utility is based on deductive logic, other researchers argued for the need to consider actual behaviour. Prospect theory builds on utility theory but adds loss aversion as a separate consideration, contending that individuals do not value similar increases and decreases of utility in the same way. In addition, probability becomes prospect, and there is recognition that subjective decision weights are attached to probabilities. In migration, individuals have to weigh the probable outcomes of migrating versus non migration, or of migrating to alternative destinations. So the migration decision depends not only on the ability to estimate probabilities but also the values of the outcomes (utility theory) and the fact that a certain level of return or risk is not always valued the same in all circumstances. For example, prospect theory would look at the loss of current income as an important consideration in return for securing a risky income abroad. There is evidence to substantiate this, and Allen and Eaton (2005: 15) have argued that 'our analysis suggests an extremely large volume of migration flows if agents are risk-neutral. Our analysis also shows that risk aversion substantially reduces these flows'. Prospect theory does not explain why there are different weightings, but it does acknowledge that they influence responses to the risks attached to outcomes.

In summary, the expected value concept considers a fully rational individual who evaluates potential outcomes according to his or her objective values and probabilities. The expected utility theory is more realistic in recognizing the existence of risk attitudes. These are expressed in the curvature of the utility function, whereby an additional parameter reflects the diverse risk attitudes of decision-makers. Prospect theory considers two further psychological traits: loss aversion and the nonlinear transformation of objective probabilities to decision weights. These theories represent steps towards a more realistic theoretical framing of decision making, but they tread a narrow pathway, even though this does broaden out from the reductionist assumptions of neoclassical approaches, to the assumptions of bounded reality.

RISK IN ECONOMIC THOUGHT

At one level, decision making—including that by migrants—is obvious and simple. The migrant first collects information, processes it and forms judgements, and these become the grounds for decision making. This simple model can be expanded into further stages to incorporate, for example, the decisions about whether to go or stay, the destination, or the timing of the migration, but the essential process remains the same. In reality we know that different individuals will emerge with different judgements at the end of this process. That may stem from the way they collected and evaluated information, or from personal preferences and contingent factors that influenced their judgements and decisions. Faced with this complex process and subjective decision making, most science disciplines employ reductionist principles to understand complex behaviour by economic agents. This is epitomised by the economists' neoclassical approach which aims to identify simple, but universal, principles underlying human decisions. While it is not claimed that these principles can explain the entirety of human behaviour, the approach does assert that they can identify some of the central determinants of decision making. The simplest models are essentially those in the neoclassical mode which assumes that individuals are maximisers, acting with perfect information. Before proceeding to consider the neoclassical approach to migration and risk, it is first necessary to place this in context of the evolution of economic thought relating to utility and expectations.

The theoretical roots of the neoclassical approach lie in the work of John Bentham (1748–1832), the founder of utilitarianism. Bentham thought that individuals made decisions on the basis of individual evaluations of the contribution of outcomes to maximizing individual utility. In 1789 Bentham introduced the concept of individual 'hedonic calculation', arguing that all individuals try to maximise their individual utilities and equate the usefulness of outcomes with the production of pleasure and avoidance of pain.

Economics turned increasingly to the natural sciences for ways to undertake such measurement. This required the assumption that individuals act in *rational and predictable* ways. These assumptions about human rationality and maximising utilities are necessary axioms for the creation of equilibrium theories by Léon Walras, Vilfredo Pareto and others in the nineteenth century, and Kenneth Arrow and Gérard Debreu in the 1950s. The price to pay for this scientific approach was the neglect of most psychological considerations in microeconomics (Kahneman and Thaler 2006: 221–222).

According to Bentham and the marginalist school, individuals know the actual utility of a particular outcome, such as migration to one country rather than another. As such, there is little or no room for risk and uncertainty in such a formulation. In contrast, expected utility is calculated as the product of probability and the value of the outcome. Typically this is reduced to a function whereby expected utility is calculated as the product of probability and the value of the outcome. In other words, the migrant weights the expected wages, or other returns, in a destination by the probability of being able to realise these. It remains an objective rational calculation, but it does acknowledge the existence of risk, or that outcomes are known only in terms of probabilities rather than definitively.

Of course, the existence of psychological determinants of human decisions was known at this time. Indeed, the founder of economic science, Adam Smith (1723–1790), had written about this and indeed he considered his most influential work to be *Theory of Moral Sentiments* (1759). Smith assumed that the study of psychology is crucial for understanding economic decisions, and painted portraits of self-interested and self-commanded economic agents. He was well aware of the limits of individual judgement and the narrowness of individual comprehension, the care of personal happiness, of family, friends, and country: 'But though we are . . . endowed with a very strong desire of those ends, it has been entrusted to the slow and uncertain determinations of our reason to find out the proper means of bringing them about' (1759: 215). Smith also noted different perceptions of gains and losses: 'We suffer more . . . when we fall from a better to a worse situation, than we ever enjoy when we rise from a worse to a better' (1759: -192). Adam Smith had pointed to the principle of *loss aversion* some two centuries before Daniel Kahneman and Amos Tversky addressed this in their founding work on behavioural economics, which is discussed later in this chapter.

Psychological considerations, however, were difficult to integrate into these formal (and very elegant) neoclassical models. Microeconomics theory assumed rational expectations by economic agents and efficient markets. In contrast, the rational expectation theory realised these limitations and did not insist that all economic agents were rational. Some agents—migrants in their choice of country, perhaps—may actually act irrationally, but each in a different way. In this case, outcomes based on irrational and/or suboptimal decisions cancel each other, and do not provide for *systemic market bias* so that equilibriums can be computed. It may not be possible to explain the decision of every migrant or non-migrant, but it is possible to explain migration in aggregate.

An important early contribution came from Daniel Bernoulli, and especially the famous St. Petersburg Paradox, which established that apparently rational individuals would decline a gamble with an apparently infinite payoff. He argued that people do not consider the expected values of bets in each round of the game, but their *expected utilities*: specifically, he referred to '*moral expectations*' by economic agents and the '*moral value*' of money (Bernoulli 1986: 24, 34). Bernoulli was the first to realise the principle of decreasing marginal utility. Bernoulli held that loss of basic goods is more significant for an individual than the loss of luxury goods. The means of covering the basic needs of life therefore have higher marginal utility than means designed for display of social status. Applying these ideas to migration, most potential migrants would stay at home because the additional income earned brings them relatively small utility (diminishing marginal utility). This applies particularly to those whose basic needs are already satisfied. They may desire luxuries, and these provide utility, but a diminishing marginal utility, so they will be unwilling to take too much risk to secure this via migration. However, if they were deeply impoverished or desperate in some other way, then migration would represent a way to secure basic goods, and hence greater utility. In other words, there would be much higher migration rates in a world of linear marginal utility changes than is encountered in the world of diminishing marginal utility. This is an important basis of the microeconomics of decision making, and most migration theories make reference to this, including Sjastaad's (1962) work on human capital and migration.

Another important idea advanced by Bernoulli to explain the St. Petersburg Paradox was the nonlinear weighting of probabilities. Individuals' fear of losses means that, as they achieve higher levels of marginal utility, they decide to stop gambling at certain thresholds. This can be extended to potential migrants who, if they already have a reasonably high level of marginal utility, will be deterred by the fear of losing what they currently have—that is, by loss aversion. Other ideas he proposed have since been questioned. This reflects the fact that his work was theoretical rather than empirical, a failing that would be addressed in the twentieth century by Kahneman and Tversky's (1979) prospect theory. Even so, his work represented an important step towards a more realistic understanding of utility.

It took over two centuries to unify all the above-mentioned concepts (marginal and expected utility, diminishing marginal utility, and nonlinear weighting of probabilities) into a single theory. Von Neumann and Oskar Morgenstern (1944) unified the concepts of marginal utility and diminishing marginal utility into the *expected utility theory*. Of particular interest to this book, the expected utility theory marries the concepts of rationality and risk aversion. According to von Neumann and Morgenstern (1944), individuals frame their economic decisions in terms of gambles and construct their utility functions on the basis of: (i) the magnitudes of potential outcomes, (ii) the probabilities of potential outcomes, and (iii) *risk* preferences.

The inclusion of risk preferences added significantly to the explanatory power of expected utility theory compared to expected value theory. The expected utility theory explained why a decision-maker may decline an outcome with the highest expected value and still maximise individual utility in a rational way. His/her preferences are represented by a utility function. If preferences are maintained in a consistent way, and the preferred alternative is available, the individual always picks this alternative. The migrant therefore would select the option which offered the highest level of expected utility. The key point is that risk attitudes are reflected in the marginal utilities.

- Risk-averse individuals have decreasing marginal utility of wealth (each additional unit of wealth brings a lower quantity of pleasure)
- Risk-seeking persons have increasing marginal utility of wealth (each additional unit of wealth brings higher quantity of pleasure)
- Risk-neutral persons have constant marginal utility of wealth (each additional unit of wealth brings the same quantity of pleasure).

Box 2.1 explains the different marginal utility functions of these three idealised decision makers. The von Neumann-Morgenstern utility function

(1944) is given by (a) the upward slope and (b) the curvature. The upward slope reflects increasing utility from an increasing wealth ('the more the better'). The curvature of the function defines whether the individual is risk averse, risk neutral or risk seeking. The curvature of the utility function in fact defines marginal utility derived from wealth.

Box 2.1   Expected Utility for Risk-Averse, Risk-Neutral and Risk-Seeking Migrants

Assume there are three potential migrants. All have an initial wealth (IW) of £3K (where K = £10000 pounds) and are offered the opportunity to take the risk of migrating. Let's assume that it is known that there is a 50% chance of the migration being successful, and the migrant will secure gains from working abroad of £2K, so that his/her wealth increases to £5K. If he/she fails to get a job abroad, then because of the costs of migrating, a loss of £2K will be incurred and his/her wealth decreases to £1K. The probability of being a winner or a loser is the same, 50% (0.5). The value function for migration is defined by curve x^α. The value of the α coefficient implies whether the individual is risk averse, risk neutral or risk seeking.

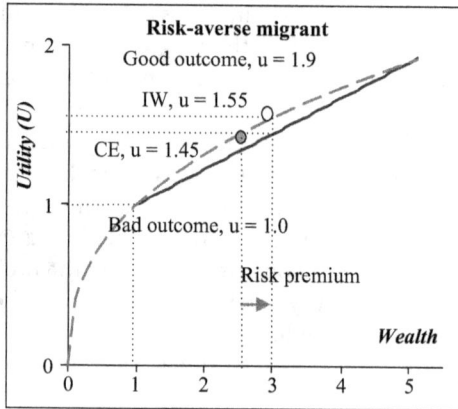

Figure 2.1   Expected utility for risk-averse migrants

When $\alpha < 1$, the individual is risk averse. The risk-averse individual, for example, has a utility function $U(w) = x^{0.4}$. His initial wealth had a utility $3^{0.4} = 1.55$. If he decides not to migrate, his utility remains 1.55. The expected utility of migration is $0.5U(1^{0.4}) + 0.5U(5^{0.4}) = 1.45$. The risk-averse individual therefore derives the same utility from a wealth of £2.6K as the risk-neutral individual would from a wealth of £3K. The difference between £3.0K and £2.6K (£0.4K) represents the risk premium that the risk-averse individual would demand for taking the risks associated with migration. In other words, his/her certainty equivalent (CE = 2.6) is £0.4K lower than expected wealth (£3K).

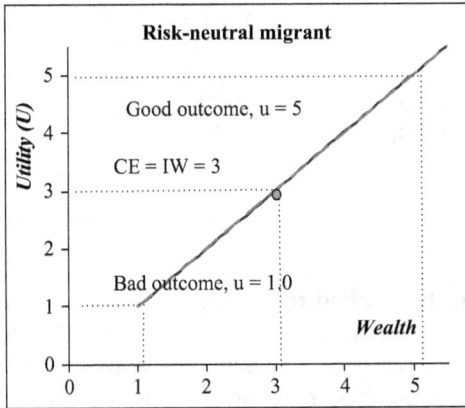

Figure 2.2 Expected utility for risk-neutral migrants

When $\alpha = 1$, the agent is risk neutral. The risk-neutral individual has a function $U(w) = x^{1.0}$. His/her initial wealth had a utility $3^{1.0} = 3$. The expected utility of migration is $0.5U(1^{1.0}) + 0.5U(5^{1.0}) = 3$. For this individual, the utility of migration is the same as the utility of his/her initial wealth. In terms of utility, he/she is neither better off nor worse off. In other words, the certainty equivalent (CE = 3) is the same as the expected wealth (£3K) and there is no risk premium to pay or demand.

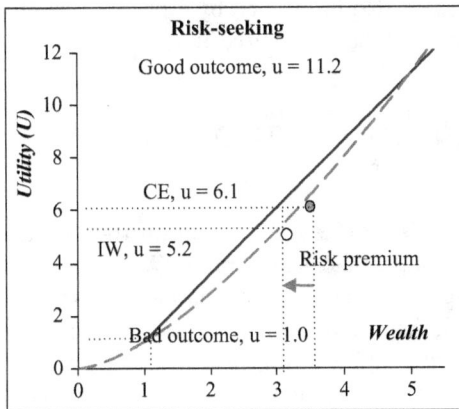

Figure 2.3 Expected utility for risk-seeking migrants

When $a > 1$, the individual is risk seeking. The risk-seeker, for example, has a utility function $U(w) = x^{1.5}$. His/her initial wealth had a utility of $3^{1.5} = 5.2$. If he/she decides not to migrate, the utility remains at 5.2. The expected utility of migration is $0.5U(1^{1.5}) + 0.5U(5^{1.5}) = 6.1$. The risk-seeker migrates, as he/she derives the same utility from wealth (£3.3K) as the risk-neutral individual from a wealth of £3K. The difference between £3K and £3.3K (-£0.3K) is the (negative) risk premium that the risk-seeker is willing to pay to migrate. In other words, his/her certainty equivalent (CE = 3.3) is £0.3K higher than the expected wealth (£3K).

Source: Authors

These ideas can be applied to migrants, with decisions about whether to migrate or stay, or the choice of destination, being influenced by whether they are risk averse, risk seeking or risk neutral. Risk aversion is the norm, and that provides an explanation of why migration is far less frequent than might be expected on the basis of, for example, wage or unemployment differentials in different locations. For example, there are only some 250 million international migrants (people residing outside their country of birth) in a global population of some seven billion. The theoretical attraction of the concept of expected utility is that differences in risk aversion/tolerance amongst individuals mean that it is rational for individuals faced with similar expected material circumstances to make different decisions with respect to migration. The same model can be extended to show the importance of initial wealth. Essentially, holding the returns from migration constant, a fully rational individual would be expected to take higher risks with increasing levels of initial wealth. These are, of course, illustrative and necessarily simplified models, and real decision making is also influenced by a plethora of other considerations, which are considered in Chapters Three and Four.

NEOCLASSICAL THEORIES, HUMAN CAPITAL, RISK AND MIGRATION

Massey's seminal review (1993) of migration theories divides neoclassical theories into two groups: macro and micro. We first consider, briefly, macro theories, but—given the focus of this chapter—mostly consider the micro, or individual, level. As noted earlier, while many of the founding thinkers recognised that individuals do not always make rational or optimal decisions, there is an underlying belief that these tend to cancel each other out, so that aggregate patterns are consistent with the assumptions of the theories of utility.

Probably the best known of the early theories within this framework was the work of Lewis (1954), Harris and Todaro (1970) and others on the role of labour migration in economic development. They essentially argued that labour would flow to areas where there were higher ratios of capital and wages, leading to long-term convergence between regional or national economies. Differentials would not, however, be completely eliminated even under idealised conditions, but would reflect 'the costs of international movement, pecuniary and psychic' (Massey *et al.* 1993: 433). We would add that these include the risks associated with migration. In these theories, wage differentials become the prime drivers of migration, and there is necessarily an exclusive focus on labour migration and labour markets.

The microeconomic models focus on individual choices (Todaro 1969; Todaro and Maruszko 1987) within a rational, maximization framework and perfect information. Individual migrants decide on whether to migrate, and where to migrate to, based on a cost-benefit calculus. This is usually modelled in terms of wages as a measure of expected utility, but data limitations have constrained the analysis of diminishing marginal utility and the estimation of probabilities. Todaro (1980: 372) assumed 'that rural-urban migration will continue so long as the "expected" urban real income (the wage times the probability of finding a job) exceeds real agricultural income at the margin—that is, potential rural migrants behave as maximisers of expected utility'. Todaro (1980) replaced the axiom of perfect labour markets by the utility function of the migrants who tried to find employment in a destination region. Instead of actual earning differentials, expected earnings entered the utility function. Freeman (1986), in his welfare state approach, pointed to the importance of differences in welfare benefits. He concluded that it was not only real direct wages, but also the benefits of the welfare state, which attracted migrants from the less developed countries and may compensate for some specific migration costs (language and culture differences, racial harassment, etc.). The approach was still based on expected utility.

Within the neoclassical framework, particular attention has been given to a strand of research known as human capital theories. Human capital theory is still centrally concerned with maximizing expected utility. The theory assumes that individuals analyse their skills and try to compute their 'net present value' with regard to the future, both in their home regions and the potential destination regions (Sjaastad 1962). Potential gains are compared with potential losses (migration costs). These costs are expressed not only in monetary terms (expenses for travel, accommodation, etc.), but also the psychological burdens (separation from family, loss of previous social networks, etc.). The migration decision is taken only when the expected returns are positive. Risks and uncertainties are acknowledged but are not explicitly quantified; instead, they are assumed to be factored into estimated future incomes and costs (Katz and Stark 1986).

The theory assumes that younger individuals with higher education can expect higher potential gains, due to their greater flexibility and lower migration costs, since they do not have such extensive social networks as individuals in middle or later life. Taking a lifetime approach enables the introduction of the discounted value of likely future earnings and the modelling of various kinds of risk/profit strategies by labour migrants. Individual migrants have an earnings function which incorporates consideration of the time-span for acquiring nationally specific human capital, such as language skills or knowledge of local practices or institutions (Chiswick 1978). Migrants initially lack such nationally specific skills and competences, and are therefore less productive and receive correspondingly lower wages than local workers with similar general educational qualifications or human capital. Over time, migrants acquire these nationally specific skills and competences, so their wages tend to converge with those of indigenous workers. Two sources of risk concern whether they can acquire such nationally specific knowledge, usually highly tacit in nature, and the time scale for their acquisition.

The model has been widely applied and elaborated by a number of researchers. Chiswick (1980), for example, argued that the personal success of particular migrants depended on the transferability of human capital between destination and generating regions, and the ability of the migrant to develop his/her social capital in the destination region. Both have attached risks which are difficult to incorporate into these models. Massey *et al.* (1993) provide a succinct summary of the human capital approach, which is redolent with notions of risk.

> Net returns in each future period are estimated by taking the observed earnings corresponding to the individual's skills in the destination country and multiplying these by the *probability* of obtaining a job there (and for illegal migrants the *likelihood* of being able to avoid deportation) to obtain 'expected destination earnings'. These expected earnings are then subtracted from those expected in the community of origin (observed earning are multiplied by the *probability* of employment) and the difference is summed over a time horizon from 0 to n, discounted by a factor that reflects the greater utility of money earned in the present than in the future. From this integrated difference the estimated costs are subtracted to yield the expected net return to migration.
>
> (Massey *et al.* 1993: 434, emphases added)

Risk in the form of probabilities of obtaining jobs, or for irregular migrants of being detected and deported, is specified in this conceptualization. However, when it comes to empirical modelling, risk is not usually explicitly analysed or estimated in most human capital research, instead being assumed to be implicit in the calculus of costs and benefits.

Clearly, human capital approaches put as much emphasis on employment prospects as they do on wages. They also conclude that the likelihood of migration is substantially influenced by the characteristics of human capital.

Individuals are more likely to migrate when they possess the type of human capital (in terms of education, experience, language competence, etc.) which means that they can expect relatively higher returns. In practice, much of the debate around this has centred on the differential returns to skilled versus unskilled migrants. The extent of the differential returns (and costs) to different human capital determines the scale of migration for each category of human capital. Puerto Rico has been the subject of considerable research on this issue with, for example, Melendez (1994) demonstrating that education and occupation were important factors: farmworkers, labourers, and craft workers were disproportionately likely to migrate from Puerto Rico to the USA, which is explained by unemployment rates in the home country, and the availability of job offers in the destination. The latter relates strongly to the notion of reducing the risk, and therefore, the costs, associated with migration. A study by Jones (1989) of the provincial-level distribution of emigration rates addressed risk issues more explicitly, if selectively, by incorporating a number of measures of both political violence and economic disruption. Political violence did influence emigration rates, but only indirectly through its impact on the economy, that is, through intermediary variables such as land disputes, strikes and disinvestment.

Migration research within the human capital theoretical framework has mostly focussed on long-term labour migration and settlement. It is far less effective when applied to other forms of economic migration, such as that by highly skilled professionals relocating between branches of an international company. Applying this framework to return migration is also problematic. Given that migrants invest in acquiring nationally specific skills in order to maximise the returns to their human capital, and that their incomes rise consequentially, the question arises of how to explain the return migration decision during rather than at the end of their working lives: wages will be either be increasing or be at their peak after a period of time, so it seems irrational to return when they can maximise the return, and compensate on the lower returns they received initially as migrants. Dustmann (1997) addressed this through considering the relationships between wages, savings and uncertainty (Box 2.2). In a later work, Dustmann and Weiss (2007) proposed three reasons why individual migrants might decide to return. First, the balance between economic returns in the destination and preferences for consumption in the country of origin may change over time. Secondly, differential wage-price ratios suggest that the individual may have higher purchasing power in his/her country of origin than in the destination. Thirdly, the value of his/her original human capital in the country of origin, and that acquired abroad, may have changed since the moment of emigration. This is especially likely in transition economies, which are subject to rapid transformation, and where there may be acute temporary shortages of particular technological, management or language skills.

Box 2.2   Return Migration, Uncertainty and Precautionary Savings

Dustmann (1997) sought to explain the migrant decision of whether and when to return to country of origin, testing a model based on wages, savings and differential labour market risks between the country of origin and destination. The model represented a significant departure from previous research in considering not so much uncertainty per se, as how uncertainty impacts on savings and the duration of migration (and therefore when to return) as joint decisions.

Migrants save more than indigenous workers because the marginal utility of wealth is increased by lower future income prospects, and by the higher marginal utility of consumption in the home country (partly due to lower prices). Uncertainty is also important: the migrant will save more if there is higher variability in income in the home country, or if his/her income is more variable than that of indigenous workers. However, the migrant may actually save less than the indigenous worker if income variability is lower in the country of origin—effectively reducing lifetime income uncertainty.

The relationship between return and uncertainty is complex, but conclusions can be drawn in some circumstances. If the wage differential between country of origin and destination is large, and there is uncertainty in the labour market in the former, then the duration of migration will increase. But if there is more uncertainty in the labour market in the destination, then the duration of migration will decrease.

Source: Dustmann (1997)

In their review of the empirical research on the application of neoclassical theories to migration, Massey *et al.* (1994: 710) concluded that there was more consistent evidence for the influence of employment-related measures, than for wages, on migration rates:

> Wage variables are occasionally found to be insignificant in migration models. . . . [But] employment variables are always significant. In the studies we have reviewed, the effects of employment-related variables generally equalled or exceeded those of wage-related indicators.

Despite the more nuanced conceptualization of human capital, this is a neoclassical approach which assumes that migration will continue until wage and employment differentials have been eliminated, that is, until equilibrium is achieved. In summary, then, although the human capital approach is useful in explaining the selectivity of migrants, it rests on unrealistic assumptions about risk, and ignores the existence of incomplete and asymmetric information. Another drawback of the theory is that it does not explain some specific types of migration flows, namely retirement migration and forced migration.

PROSPECT THEORY, RISK AND LOSS AVERSION

In his book, *Theory of Moral Sentiments* (1759), Adam Smith noted that individuals could have different perception of gains and losses: 'We suffer more . . . when we fall from a better to a worse situation, than we ever enjoy when we rise from a worse to a better' (1759: 311). In effect, Adam Smith pointed to the principle of *loss aversion* some two centuries before Kahneman and Tversky (1979) produced the first version of their prospect theory. The latter immediately attracted the attention of behavioural researchers, and in 2002 Daniel Kahneman was awarded the Nobel Prize in economics for his work on the psychology of judgment, decision making and behavioural economics. (Amos Tversky had died in 1996.)

Prospect theory overlaps with expected utility theory in many areas but fundamentally differs in being concerned with bounded rationality. Both theories considered utility to be a product of value and the *probability* of the outcome. Both also accept neoclassical assumptions about equilibrium, and that individual decision making seeks to maximise benefits, aiming to achieve optimal and/or satisficing solutions. However, prospect theory replaced the notion of 'utility' with 'prospect', which accentuates the *uncertainty* of the outcome, and the need to explain this more fully as well as the psychological factors impacting on choice. A key difference between the theories lies in the computation of potential outcomes (prospects instead of utilities), especially in their assumptions. The most important differences refer to: (i) different weights are assigned to positive and negative outcomes by the decision makers; (ii) nonlinear weights are assigned to small, medium and large probabilities; (iii) reference points (people consider changes in utility rather than their total utility); and (iv) the concept of rank-dependent utility models. Prospect theory has particular resonance for understanding migrant decision making, and how potential migrants view losses versus gains from migration, the risk of significant if low probability losses in many instances, and changes in their current well-being as a result of moving to a new location.

The expected utility theory uses a single parameter (curvature of the utility function) to compute the value of the expected utility from the probability and value of the outcome. In contrast, prospect theory includes more parameters when computing the utilities of potential outcomes (prospects): loss aversion and the nonlinear weighting objective probabilities. The advantage of this is that prospect theory provides a more general framework for explaining decision-making processes, and can address some of the cases which classical expected utility is unable to explain. In fact, expected utility theory is seen as a special case of prospect theory when: (i) there is no difference between how individuals value losses and gains, and (ii) individuals are able to accurately estimate objective probabilities and assign decision weights identical to the values of the probabilities. However, the unrealistic nature of such conditions has made prospect theory a more attractive proposition for many researchers.

Expected utility theory assumes that individuals treat gains and losses equally. In other words, a migrant would value a gain of £1000 pounds and a loss of £1000 pounds as having the same utility, but with different signs. In contrast, and drawing on empirical evidence, prospect theory assumes that the psychological costs of losses are on average about 2.25 times higher. Kahneman and Tversky (1979) found 2.25 *a median* value of the loss aversion parameter. Applying these ideas to migration, *prospect theory asserts that because gains and losses loom differently*, the potential migrant would ask on average for a potential gain of 2.25 * 1000 = £2,250 pounds, if he/she were also faced with a potential loss of £1000 pounds as a result of migrating.

Another important feature of prospect theory is that it assumes that *decision weights* are assigned to objective probabilities in a nonlinear way. In expected utility theory, if a migrant faced two outcomes from different migration strategies, with probabilities of 0.1 and 0.9 respectively, it is assumed that the second outcome is nine times more important. In contrast, prospect theory contends that individuals overweight low probabilities, and underweight moderate and large ones. Kahneman and Tversky (1979: 267, Problems 7 and 8), illustrate this through experimental research on gambling. Overweighting low probabilities explains widespread interest in lotteries, which attract investment beyond the objective probability of winning because of the weights attached to the large if low probability prizes. It also explains the popularity, for example, of buying insurance against terrorist attacks in air travel, despite their relatively low probabilities. Underweighting of medium and large probabilities lies behind the low willingness to stop smoking or drive more safely. A migration example would be what we can term the 'streets are paved with gold' phenomenon: that is, the overweighting of the probability of very high levels of success (e.g. becoming a successful business owner) and underweighting the probabilities of incurring health costs through migration. Tunali (2000) provides evidence of this in the case of Turkish migrants: many migrants had expectations of negative returns, but there was a willingness to invest in migration with a high probability of such returns because of the potential for a very low probability but very large payoff.

Similarly to expected utility theory, prospect theory proposes that guaranteed or risk-free gains are considered more valuable than risky ones by individual decision-makers. Kahneman and Tversky (1979: 267) demonstrated this through experimental research where individuals were faced with imaginary choices between different holidays, with different values and probabilities. These findings can be projected onto an imaginary potential migrant who is faced with two choices, each of which constitutes two options. In the first choice, friend A has found a guaranteed job but with relatively low pay, while friend B has found a much higher paying job but with only a 50% chance that this will be obtained.

Extending Kahneman and Tversky's findings, most migrants would be expected to choose the former option, that is, certainty. However, if the second choice was between a 1% chance of a high wage, and a 2% chance of a lower wage (i.e. the probabilities are still in a ratio of two to one), then, extending Kahneman and Tversky's findings, most individuals would choose the first option, demonstrating the nonlinear weightings of probabilities.

The expected utility theory also assumes that people frame their decisions in terms of gambles and that the total wealth of an individual is the starting point for his/her valuation of different outcomes. In contrast, prospect theory assumes that gambles are framed only in terms of gains and losses, based on the assumption that the individual relatively quickly adapts to a certain level of utility. This level is considered the *reference point*, and it is the absolute changes rather than total wealth which are critical (Kahneman and Tversky 1979: 277). A simple application of this principle to migration would be to compare two migrants. The first is a relatively wealthy, skilled migrant to the USA who fails to find a job and, therefore, effectively has a negative return equal to the travel costs and the opportunity costs of the job left in the home country. The second is an unskilled migrant who goes to the USA and secures a low paying job but with a salary that is higher than he received in the country of origin. The skilled migrant still has far more wealth, but the unskilled migrant values the outcome more highly because he or she has different reference points.

Decision weights should not be confused with subjective probabilities. Subjective probabilities refer to incorrect estimates of objective probabilities. According to Kahneman and Tversky (1979: 280), the 'decision weights measure the impact of events on the desirability of prospects, and not merely the perceived likelihood of these events'. Tversky and Koehler (1994) asked a sample of Stanford University students to estimate the probability of unnatural causes of deaths in the USA (fatalities, murders, manslaughters and other unnatural cases). The students estimated that the probability of death by unnatural causes is 53%, while the correct value is 8%. The authors of this book repeated the same experiment with students at the University of Economics in Bratislava (Slovakia). The Slovak students estimated the likelihood (i.e. subjective probabilities) of unnatural deaths to be 27%, while the correct value was 6%. Murders and fatalities loom larger to individuals than do the standard causes of death (e.g. diabetes). They attract the attention of individuals and their likelihood is overestimated. Applying these ideas to migration, for a migrant who 'really wants to go to the USA', the positive outcomes are likely to loom large. The probability distribution from the expected utility theory has been replaced by the *weighting function* in prospect theory (Box 2.3). The weighting function describes the transformation of objective probabilities to decision weights.

Box 2.3   The Value Function and the Weighting Function

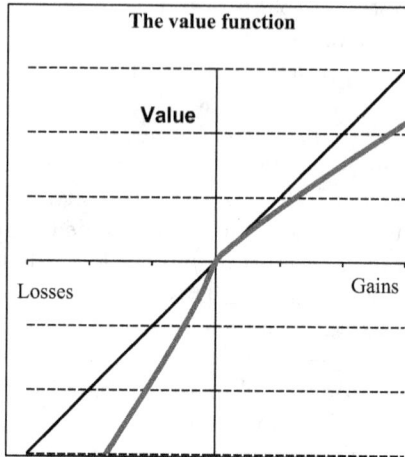

Figure 2.4   The value function

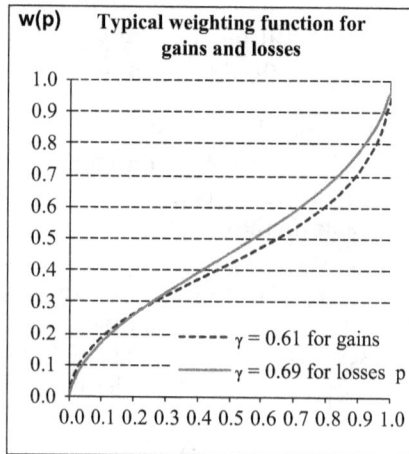

Figure 2.5   Typical weighting function for gains and losses

The *value function* describes the assigning of values to potential outcomes by decision-makers. The asymmetric curvature of the function reflects higher sensitivity to losses than to gains. The *weighting function* describes the over-weighting of small, and the underweighting of medium and large, probabilities by decision-makers.

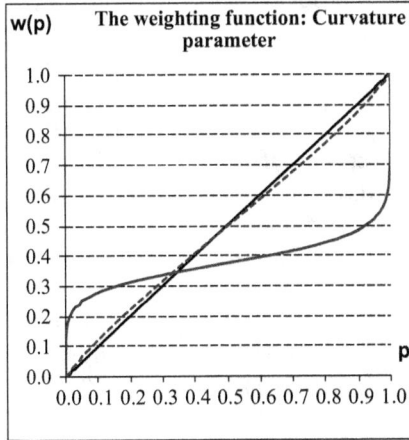

Figure 2.6   The weighting function: curvature parameter

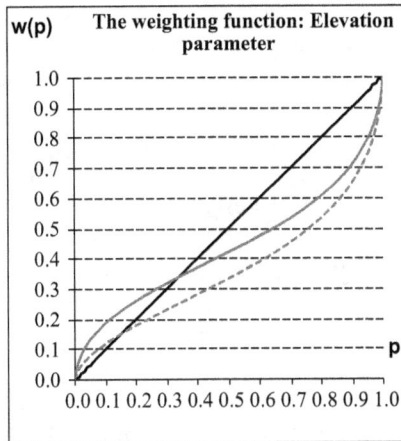

Figure 2.7   The weighting function: elevation parameter

Discriminability and curvature: The dotted line indicates more accurate weighting and approaches a straight line; the step-like solid line indicates low ability to assign decision weights.

Attractiveness and elevation: The dotted line indicates lower levels of over-weighting small probabilities and lower attractiveness; the solid line is more elevated and indicates an attractive option and overweighting of small probabilities.

Sources: Authors, based on Tversky and Kahneman (1992), and Gonzales and Wu (1999)

Gonzales and Wu (1999) identified other properties of the weighting function. They considered that weighting may incorporate two independent, yet intertwined, psychological traits: (a) discriminability, and (b) attractiveness (Box 2.3). Attractiveness is of particular relevance, referring to interpersonal (and also intrapersonal) differences in the assigning of decision weights to certain outcomes. If one risk domain seems more attractive than another one—as in our example of the migrant who really wants to go to the USA—then that domain will receive a higher decision weight. The more an individual wishes for a certain outcome, the higher the weights that are assigned and the higher elevation of the weighting function. The two traits, however, are inter-related. The more attractive the outcome, the more likely it is to generate interest by the decision-maker. As Gonzales and Wu (1999: 161) comment: 'We like what we know and we know what we like'.

Prospect theory is central to Kahneman and Tversky's work and has generated considerable discussion and controversy. Most research has aimed to provide realistic estimates of parameters in the value and weighting functions. Table 2.1 summarises several studies in this field. Most compare the certainty equivalents and actual values of outcomes, usually in context of a series of hypothetical gambles, in order to establish the parameters of the value and weighting functions. Their findings indicate that:

a) Most estimates of the value function parameter (the curvature of the utility function) lie in the interval 0.52–1.00, indicating that individuals are risk averse;
b) Most estimates of the weighting function parameter lie in the interval 0.60–0.71 for gains and 0.51–0.76 for losses, indicating that the decision weights tend to diverge significantly from the objective probabilities.

There are some real-life studies confirming the existence of the phenomena assumed by prospect theory, namely loss aversion and nonlinear weighting probabilities, including: Nicolau (2010) for airline demand, Betts and Taran (2006) for used car markets, Camerer *et al.* (1997) for taxi drivers, and Zhang and Semmler (2009) and Barberis *et al.* (2001) for financial markets.

Amongst these various studies, the authors provide one of the few studies to have applied prospect theory to migration (see also Tunali 2000; and Allen and Eaton 2005 specifically on loss aversion). They applied prospect theory to migration studies in two studies undertaken in Slovakia. The Slovak Republic is a less developed EU and OECD member country, with a median net monthly wage of 514 Euros in 2012. The first experiment, with a sample of 19 construction managers, was performed in the field, while the second experiment with a sample of 96 university students was undertaken in a controlled laboratory environment (Table 2.2).

Table 2.1   Selected studies estimating parameters of the value and weighting functions in prospect theory

Source	Value function (α)	Weighting function (γ)	Loss aversion (λ)	Sample character and size
Tversky and Kahneman (1992)	0.88	0.61	2.25	25 students
Tversky and Fox (1995)	0.88	(0.69)	×	141 students
Wu and Gonzales (1996)	0.52	0.71 (0.68)	×	420 students
Gonzales and Wu (1999)	0.49	0.68	×	10 students of psychology
Abdellaoui (2000)	0.89	0.60 (0.60)	×	64 students
Bleichrodt and Pinto (2000)	0.77	0.67 (0.55)	×	51 students of economics
Kilka and Weber (2001)	0.76–1.00	(0.30–0.51)	×	55 students of finance
Abdellaoui *et al.* (2003)	0.91	(0.76)	×	52 students
Stott (2006)	0.19	0.96	×	96 students
Abdellaoui *et al.* (2008)	0.86	0.60	2.61	48 students of mathematics and psychology
Abdellaoui *et al.* (2011)	0.79	0.73	2.47	48 students of management
Baláž *et al.* (2013)	1.02	0.58	2.50	19 construction managers
Baláž *et al.* (2013)	1.00	0.62	2.10	96 students of economic and psychology

Notes: Value of the parameter of the weighting function for losses in parenthesis.

Source: Authors

In sample one (construction managers), the overall findings generated values of risk aversion parameter of 1.021 for the value function and 0.571 for the weighting function. The median value of the loss aversion parameter (2500) was quite close to that produced by Tversky and Kahneman (1992). The sample was dominated by young and relatively well-paid men with above-average incomes, and this may explain the somewhat higher value function compared to Tversky and Kahneman. Seven of the construction managers had previously been migrants and had worked abroad. Migrants tended to have relatively higher levels of risk aversion: the median value of the parameter was 1.015 for migrants and 1.045 for non-migrants. The migrants, however,

Table 2.2 Estimates of the prospect theory parameter values for those with and without migration experience in two Slovak samples

Data type	19 construction managers in Slovakia			96 students in Bratislava		
	Median value	Individual-level median	Individual-level mean	Median value	Individual-level median	Individual-level mean
All subjects						
Value function (α)	1.021 (0.008)	1.035	1.042 (0.059)	1.003 (0.005)	1.016	1.027 (0.046)
Weighting function (γ)	0.578 (0.040)	0.571	0.541 (0.202)	0.623 (0.033)	0.499	0.578 (0.423)
Loss aversion (λ)	2.500	2.300	2.584 (1.063)	2.100	2.300	2.404 (1.021)
Migrants						
Value function (α)	1.015 (0.009)	1.015	1.045 (0.066)	1.002 (0.004)	1.016	1.033 (0.051)
Weighting function (γ)	0.660 (0.064)	0.621	0.594 (0.203)	0.681 (0.039)	0.597	0.605 (0.422)
Loss aversion (λ)	2.233	2.300	2.586 (1.275)	2.300	2.700	2.514 (1.253)
Non-migrants						
Value function (α)	1.018 (0.011)	1.045	1.040 (0.579)	1.001 (0.005)	1.016	1.026 (0.045)
Weighting function (γ)	0.529 (0.041)	0.478	0.511 (0.204)	0.601 (0.038)	0.485	0.571 (0.426)
Loss aversion (λ)	2.500	2.400	2.583 (0.982)	2.100	2.300	2.377 (0.936)

Notes: Value of the parameter of the weighting function for losses in parenthesis.
Source: Baláž et al. (2013)

had slightly lower levels of loss aversion: the median value of this parameter was 2.3 for migrants and 2.4 for non-migrants. The greatest differences between the two groups were in the median values of the weighting function: 0.621 for migrants and 0.478 for non-migrants. Migrants seemed to be better in setting decision weights closer to objective probabilities in transforming probabilities to decision weights, and we can speculate that this may be due to their greater life experiences. The sample has the advantage of not being based on students, like most of the other research reported here, but the sample was very small. In summary, the results were similar to those of Tversky and Kahneman except for the coefficient of the value function, which was slightly higher than 1, indicating the overall sample was risk seeking. The reason for this may be that the sample is mostly made up of young men who tend to be overconfident, and also considered the hypothetical gambling sums to be relatively small. This sample was very small, and it was not possible to test for significant differences; it effectively represented a pilot study, albeit on a particularly interesting group of respondents.

Most of the findings from the first sample were replicated in the second experiment involving a larger group of student respondents. The latter generated values $\alpha = 1.003$ (for risk aversion), $\gamma = 0.623$ (for weighting function) and $\lambda = 2.100$ (for loss aversion), which were also quite close to those estimated by Tversky and Kahneman (1992). The median sample values for particular population groups indicated that men were somewhat more risk seeking than women (the respective values of the α parameter were 1.007 and 1.002) and also had higher discriminating probabilities (the respective values of the parameter were 0.626 and 0.480). Of particular interest for this book, migrants had somewhat higher levels of loss aversion. The median value of the λ parameter was 2.300 for migrants and 2.100 for the non-migrants. This was the opposite of the smaller sample of construction workers, where migrants had slightly lower levels of loss aversion. However, the differences between the groups were not significant. The findings are somewhat surprising. Migrants have had opportunities to live and study in foreign countries, with the different learning experiences and challenges this poses. Any migration is a risky undertaking with uncertain outcomes. The authors had initially assumed that migrants would, therefore, have had more training in and greater ability to estimate probabilities, but if that is so then such abilities may be specific to migration risks, and were not discernible in the more generic experiments they participated in.

Prospect theory has a number of detractors, of course, and one of the most recurring critiques centres on the extent to which individuals (or individual migrants) are able to calculate probabilities, risks and returns. There have been several theoretical responses to this critique by those who adhere to the notion that probabilities can be calculated, however imperfectly. One of the most promising responses is the priority heuristic which, by application to migration, suggests that migrants would focus on only a relatively narrow set of criteria and would use lexicographic rules to evaluate the often massive and confusing evidence available to them about potential destinations. Another response is the transfer of attention exchange models which emphasise the importance of how decision-making choices are configured—and the possibility that different decisions will be made in different circumstances. In view of these limitations, there is a need to consider alternative theoretical perspectives. One radical alternative is nonprobabilistic theories which suggest that individual choices may be made under very different types of procedures, including fuzzy logic; this recognises that there may be degrees of truth and that individuals may settle for acceptable levels of likely error in their decisions. These approaches are considered in Chapter Three.

CONCLUSIONS

The neoclassical approaches provide a major perspective on economic migration and mobility driven by income differences, and at various points does partly engage with notions of risk. Economic migration is probably the most common type for migration flows and, due to the availability of relatively good statistical data and its policy importance, neoclassical concepts have frequently been modelled and tested. There are, however, limitations to the contribution of neoclassical theories of migration. The most obvious is the assumption of *Homo economicus* and the neglect of the noneconomic determinants of migration behaviour, which is especially significant with respect to noneconomic migration. The basic neoclassical framework also assumes that individuals make a rational cost-benefit analysis of the expected discounted returns of migration. The decision to migrate is taken only when the gains expected from migration (sum of expected utilities) is greater than the expected losses (Chiswick 1980; Mincer 1978). Losses are associated with migration costs and mostly include the costs of travel and relocation, and the opportunity costs of time. Some neoclassical researchers do acknowledge the importance of psychological costs, but these costs are difficult to quantify. The neoclassical approaches also treat gains and losses equally, assuming that losses from migration loom with the same magnitude in migrants' minds as gains, but having a different sign.

Neoclassical theories refer to the expected utility framework, but draw on it selectively. Decisions taken under the expected utility theory, in the theories of von Neumann and Morgenstern (1944), have three major components that explain utility maximization: the values of the expected outcomes, the probabilities of outcomes, and the risk preferences (shaping the curvature of the utility function). In migration research, it is the values (in terms of income differences and net gains from migration) which have attracted by far the largest attention. Far less is known about how migrants compute the probabilities (and therefore the risk) of getting an expected job and the income associated with it. No migration study has tried to compute the expected utility via risk preferences. Yet the importance of risk is recognised by some of the key exponents of human capital theory approaches, such as Chiswick (2008: 64):

> It is often said that immigrants are different from the people that they leave behind in the origin and the people they join in the destination. They are sometimes described as more aggressive, risk taking, forward looking, and avaricious or entrepreneurial, and sometimes as healthier.

This lacuna is surprising given the prevalence of risk aversion among individuals and the importance of risk attitudes in expected utility theories.

Prospect theory offered one major response to the limitations of expected utility theories. In particular, it recognised that individuals assign different weights to positive and negative outcomes (they are loss averse, assigning greater weights to losses), and assign nonlinear weights to probabilities (overestimating small probabilities). Prospect theory has particular resonance for understanding migrant decision making, especially how potential migrants view losses (perhaps an existing job, social network, etc.) versus gains, and the positive risk of significant if low probability gains (the lure of 'the streets are paved with gold'). However, there has been no research to date, to the authors' knowledge, which has sought to apply prospect theory to an empirical analysis of migration.

Prospect theory, although an advance on neoclassical theory in recognizing bounded rationality, still has significant weaknesses. It is a rational decision-making model, which assumes that perfect information is available to individuals who are able to process this. It does not take into account either the differences in the extent to which individuals are risk tolerant/averse, or how they operate in the face of either limited information, or complexity and information overload. These themes are taken up in the following chapters.

3   Behavioural Economics Approaches

INTRODUCTION

The previous chapter mostly focussed on variants of utility maximising mod-els, and individuals as rational decision-makers. Although such approaches provide insights into fundamental economic relationships that can shape migration, they have limitations. As Kahneman and Thaler (2006: 221) argue in a general context, but with application to migration:

> The assumption that utility is always maximized allows often surprising inferences about the nature of the desires that guide people's ever-rational choices. This methodology has had many uses and undeniably has charm for economists, but it rests on the shaky foundation of an implausible and untested assumption.

Arguably, most people do aim to maximise their utility, or at least to achieve satisfactory levels. However, this is problematic because utility max-imisation requires forecasting a number of possible outcomes. As Kahneman and Thaler (2006) argue: 'People do not always know what they will like', and may make systematic errors in predicting utility from future outcomes. They contend that it is not so much that people do not know what they like, but that they do not know what they will like in the future—and this can be especially challenging when there is a considerable gap between their pres-ent and future circumstances. This is very much the case with migration, which may transverse borders between economic, cultural and regulatory systems that significantly shape future outcomes.

There are four main reasons why errors occur in hedonic forecasting (Kahneman and Thaler 2006): changes in your emotional state in future; because the decision has not focussed on the most relevant aspects of the outcome; flawed understanding of previous experiences; and failure to predict future adjustment to new life circumstances. All four apply to migration: being away from a familiar environment and under stress may change your emotional state; you may have focused on say wages, and ignored the importance of identity; you may have misread your proven ability to adjust to the demands of intranational migration to imply that you will cope effectively with international migration; and you may have underestimated how well you will adjust to living in a different country.

Behavioural economists, such as Kahneman (1994), have sought to explore the limits of rational decision making: specifically, to understand how individuals make decisions faced with imperfect information, their imperfect ability to process and analyse this, as well as the role of values and preferences, and the weighting of information. This has been taken up in a number of fields, especially health and finance, but has been surprisingly neglected in migration research, with a few exceptions discussed later. It is particularly surprising in view of the prominence of risk in migration, as discussed in Chapter One. The reasons are complex but include secondary data limitations, and the difficulties of undertaking complex survey work with specific, and often difficult to engage, groups such as migrants. It may also reflect a relative shift to qualitative research in migration studies in recent years. There may also be a predisposition in migration studies to view migration as a unique form of behaviour and necessarily reliant on migration-specific theories. In contrast, behavioural economics views migration as one type of risky behaviour amongst many others—such as investment behaviour, selecting employment, or participating in risky sports—which should be considered within the same broad theoretical approaches as other risky behaviours.

RISK, UNCERTAINTY AND PROBABILITY

The concepts of risk and uncertainty (ambiguity) are used interchangeably in many areas of social science, including migration studies. Both refer to a decision-maker who faces unclear outcomes of his/her decision, yet they are also fundamentally different, at least in theoretical terms. The economist Frank Knight (1921) was the first to clearly distinguish the two concepts. Knight defined risk as 'measurable uncertainty' or 'known unknowns' while uncertainty refers to 'unknown unknowns'. In real life, 'computable' risks are quite rare and an overwhelming majority of decisions are made under uncertainty, or—as discussed in Chapter One—in terms of very broad, qualitative expectations.

Knight's distinction is of course a positivist one where risk is understood as objective and measureable, and this stands in stark contrast to the social constructionist and emotive conceptualisations considered later in the book. For postmodernists in particular, risk, uncertainty and knowledge have blurred and shifting meanings (see Chapter Six). However, even those working within a positivist framework also find that, outside the casino and economic psychology laboratory, there is no clearly defined border between risk and uncertainty. These debates are translatable into migration decision making—do individuals act in the face of uncertainty about the destinations, or with some knowledge of probabilities (e.g. 8 out of the 10 people from my village found a job within a year), or at least with some broad impressions of probable outcomes? In reality, their decisions—based on a number of considerations, ranging from the economic to the cultural, and emotive—are likely to involve a high level of uncertainty, ranging from weak uncertainty (blurring into generalised estimates of future outcomes) to severe uncertainty. People generally dislike uncertainty, and usually try to convert it to risk, as is illustrated by the classic five urn problem outlined in Box 3.1. 'Knowledge' of the probabilities provides a feeling of perceived competence, which exerts a powerful influence on decision making.

Box 3.1   Measuring Risk and Uncertainty in Behavioural Economics

Most of the work on risk versus uncertainty in behavioural economic is based on experimental research which simulates a number of hypothetical (and occasionally real) gambles. The five urn experiment is one of the best known examples. In Urn 1 the decision-maker is faced with risk, that is, known risks, but in all the other urns faces different degrees of uncertainties, which climaxes in total uncertainty in Urn 5.

- Urn 1 contains 100 balls, of which 50 are white and 50 are black.
- Urn 2 contains some balls. Neither the numbers nor colours of the balls (they may be black, white, yellow, green, etc.) are known. However, there is information that somebody drew some of the balls, and about half were white on that occasion.
- Urn 3 contains 100 balls; some are black, some white, but in unknown proportions.
- Urn 4 contains an unknown number of balls. It is known there are five colours, including white, but the proportions of the different colours are unknown.
- Urn 5 is a riddle. It is not known if there are any balls at all, let alone the number of colours and their proportions.

If a white ball is drawn, a prize of 100 pounds is won. Which urn would the participant prefer to draw from? Most people would prefer Urn 1, but does this offer the highest chance to win? In fact, we are unable to say which urn provides the highest probability to win. Urn 1 provides a probability of 0.5 to win. Urn 2 may seem the second best candidate, because each second ball drawn has been white so far. In fact, this is based on a small sample, and the actual proportions could be better or worse. Urn 2 may contain as many as one million balls and, by chance, all the white balls have already been drawn. As for Urn 3, the probability of drawing a white ball is in the interval from 0.01 to 0.99. Urns 4 and 5 provide almost infinite probabilities for a white ball to be drawn. It may actually happen that Urn 5 contains white balls only, no white balls, or even no balls at all. Lacking knowledge of the probability distributions in Urns 4 and 5, however, is a major impetus for most people to prefer Urn 1, followed by Urns 2 and 3.

Source: Authors' interpretation of a range of experiments from behavioural economics.

The urn problem also indicates that 'probability' may be a subjective term and depends on the personality and beliefs of the observer. T. Bayes and R. Price (1763) identified probability with the *degree of belief* that some event will or will not happen. Essentially, they asserted that when a certain event is observed, its probability distribution can be inferred from the observed outcomes. Bayesian probability is also based on the notion of the state of knowledge. When considering a new event, an individual begins by considering some prior probability, which is then updated as new data are collected. As an example, take Urn 4, where there are balls of five colours whose proportions are unknown (Box 3.1). What is the probability that the white ball is drawn? Most observers would probably use the principle of indifference: as there are five possible outcomes, the prior probability of each outcome is considered to be 0.2, or one in five. However, this assessment will be updated in the light of new data. For example, if 900 of the first 1000 balls drawn were white, then the subjective belief in the probability of drawing a white ball would increase to 0.9: this is the posterior probability.

The Bayesian concept of probability has two forms. The 'subjectivist' form contends that the state of knowledge measures nothing more than the observer's strength of belief. The objectivist form takes a more positivist view and assumes that if two or more people have the same information about a certain proposition, apply consistent and rational methods of measurement and compute the same probability of the event, then the probability is objective. However, observation-based probability says nothing about strength of proof. In the example in Box 3.1, the available evidence for both Urns 1 and 2 indicates a Bayesian estimate that 50% of the balls may be white. However, the degree of certainty/uncertainty is different, which is why most participants prefer the first urn.

These deliberations may seem rather abstract but they can be applied to migration. For example, a prospective migrant may be considering three destinations, the USA and two other countries. In the USA, in addition to a quota system, 50,000 green cards are issued annually to people from all over the world based on a random lottery draw. The prospective migrant is able to check the average numbers of applicants in recent years and compute the probabilities of lottery success. This is equivalent to Urn 1 in Box 3.1. In contrast, a second country may have a visa system with no quotas and no lottery programme, and the number of applicants is unknown, but lists of visa decisions issued each year are known and about one half of published decisions on visa applications are positive. This is effectively Urn 2: total numbers of applications are not known, and as the authorities may have not made formal decisions on many of these (dismissing them as incomplete, or on other grounds) the list of decisions, although known, has limited certainty. Finally, a third country may not disclose either the number of applicants or the numbers of visas granted, and this would represent Urn 5.

As demonstrated above, there are several ways for converting uncertainty to risk. The question is whether individuals make decisions in the same way in the face of uncertainty as in the face of risk. The most influential work in this area is Savage's (1954) theory of subjective expected utility. He created an axiomatic system that extended expected utility under conditions of risk to conditions of uncertainty. Under risk conditions, the probabilities of events are known. Decision outcomes are evaluated according to their utilities and probabilities. Under uncertainty conditions, the probabilities of outcomes are unknown and the outcomes of decisions depend on which event happens or not, rather than objective or known probabilities. If the axioms refer to similar events, Savage showed that the events may be analysed using a similar computational framework. Savage's system therefore provided an extension of the expected utility theory from risk to uncertainty. Empirical investigations have indicated that the extension was reasonable, assuming some reservations and limitations. In fact, most individuals eschew uncertainty and wherever possible try to convert uncertainty to risk.

The most important axiom in Savage's system is the 'Sure Thing Principle', which is similar to the Independence Axiom in expected utility theory. This can be illustrated by a migration example. A prospective migrant is considering visa applications to Canada (alternative A) and the United States (alternative B). If he/she gets a visa for Canada (event E_1) his/her annual income is expected to be 20,000 pounds. If the USA visa is granted (event E_2), the annual income is also expected to be 20,000 pounds. In other words, their utilities are the same. However, if the migrant believes that he/she is more likely to obtain a Canadian than a USA visa (event E_1 is more likely than event E_2), a Canadian visa will be applied for. This reflects a preference for risk over uncertainty. The simple example can be made more complex by varying the utilities and probabilities for the two migration events.

Ambiguity Aversion, Source Preference and Source Sensitivity

Soon after the publication of Savage's work on subjective expected utility, discussions started about whether the concept of expected utility accurately describes behaviour under uncertainty. Experimental economics provided evidence that decision-makers may make different decisions under risk and uncertainty conditions. Ellsberg (1961) popularised a famous paradox on decision making under risk and uncertainty. The basic form of the paradox is expressed in terms of a gamble focussing on two urns: Urn 1 contains 100 yellow and black balls, but the proportions of colours are unknown—that is, uncertainty. Urn 2 contains 50 yellow and 50 black balls—that is, known risk. The subjects in the experiment were asked to decide how much they would bet on a yellow or black ball being drawn from Urn 1 and Urn 2, with the prize being the same in each case. The subjects were willing to bet more on Urn 2, where they faced (known) risk rather than the uncertainty of Urn 1. When individuals lack clear or complete information they are unable to maximise their utilities, and maximising and optimal decisions are replaced by satisficing and suboptimal ones.

The next development was recognition that not all forms of uncertainty are the same. Fox and Tversky (1995) defined two new phenomena: source preference and source sensitivity. *Source preference* refers to preferring some sources of uncertainty to other ones. The choice between two prospects depends not only on the degree of uncertainty, but also its source. In Fox and Tversky's experimental research, their American subjects preferred betting on the weather in San Francisco to betting on the weather in Istanbul. A similar experiment in Slovakia by the authors showed that Slovak students preferred betting on Slovak, rather than Greek, inflation rates. The more severe the uncertainty, the lower the willingness to bet on an event. Slovak economics students were willing to bet more on the distribution of letters in the Slovak language than in the Greek one, although the true distribution was unknown for both languages. Fox and Tversky suggested that ambiguity aversion (Ellsberg's paradox) is a special case of source preference, where risk is preferred to uncertainty (see also Baláž *et al.* 2009).

Source sensitivity refers to decisions weights. Prospect theory (see Chapter Two) indicates that individuals are more sensitive in distinguishing an increase in probability when it: (i) makes one uncertain event more probable, and (ii) turns impossibility to possibility or a possibility to certainty. In other words, an increase in probability from 0.3 to 0.4 impacts decision weights less than increasing probability from 0.0 to 0.1 or from 0.9 to 1.0. Fox and Tversky also suggested that individuals are less sensitive to changes in uncertainty than to changes in risk. The differences in the amounts that people are willing to spend on risky versus uncertain bets are much lower under non-comparative than comparative conditions. Source sensitivity arises only under comparative conditions—when the individual directly compares the two bets—and not in independent evaluations of uncertain events.

What do source preference and source sensitivity imply for migration decision making? Prospective migrants may prefer potential migration destinations for which they also have a preference for the available information source (perhaps close friends). This can again be illustrated by the visa application example. Potential destinations with a high possibility of obtaining a visa would clearly be preferred to those with low rates of acceptance.

However, migrants are likely to consider several potential destinations so that migration decisions are formed under comparative conditions, and source sensitivity is important. Additionally, even a small change in probability could make a big difference to the migration, although this is conditional on the type of uncertainty that is faced. Thus an increase in probability from 0 to 0.1 (following the first opening up of job opportunities for migrants in a previously closed country), although still a low probability, or from 0.9 to 1 (following a massive economic boost in the destination, so all migrants are guaranteed to get jobs), which was already high, would have a far greater impact on decision weights than an increase in probability from 0.3 to 0.4.

Knowledge, Risk and Uncertainty in Migration Decisions

Potential migrants can be expected to want to decrease uncertainty, and effectively transform this into risk-like probabilities—that is, based on greater certainties, however approximately these are expressed. This requires that they engage in information search and evaluation. Information search has its costs and these constitute part of the total costs of migration (see Chapter Four).

There are a number of sources of information and knowledge that can be used to reduce uncertainty, such as websites, books and newspaper reports. However, arguably the most important are social networks, which can provide both knowledge and direct practical help with housing and finding jobs. Social networks are a valued resource for migrants (Gamburd 2000), constituting location-specific capital (de Vanzo 1981) and reducing the costs of migration (Massey *et al.* 1993). After migration by the first migrant from a particular generating region to a destination, the monetary and psychological costs of migration decrease for other members of the migrant's social network (Carrington *et al.* 1996). The probability of 'success' and 'security' increases as the size of the migrant community increases over time, while uncertainties become converted into risks, at least partially.

Both informal and formal channels are used to reduce uncertainties and risks. Informal channels have an advantage in providing potential migrants with tacit knowledge, which is difficult to acquire by other means (Williams and Baláž 2008, chapter two). Migrants not only find it easier to gather information from friends and family, compared to official information channels, but they may also place greater trust in such information. The reliance on networks can be linked to the concepts of source preference and source sensitivity. Much of the key information required by potential migrants is tacit in nature, such as how best to fill in visa applications, the costs of living at different stages of migration or the real chances of getting particular jobs. Such tacit knowledge is difficult to convert into explicit knowledge. Not surprisingly, family or co-ethnic networks are often preferred sources of tacit knowledge, as they are characterised by higher levels of trust and lower communication barriers compared to other channels. Information and support provided by social networks is also important in turning severe uncertainty to computable risk and/or near certainty about the outcomes of migration (source sensitivity). Family or kin, for example, may provide a 'guaranteed job' for the first-time migrant. For potential migrants, turning possibility to certainty via social networks may be more important than increasing the probability of obtaining better paying jobs via more extensive information search.

Reliance on networks is related to the potential migrant's perceived competence, and this can change over time, especially in response to acquiring migration experiences, which may lead to re-assessment of the importance of the network. De Jong *et al.* (1983), for example, showed that rural Filipinos, who visited Manila several times and/or had migration experience, were more likely to engage in rural-urban migration. Arguably this reflects the acquisition of mobility competences that convinced them that they can manage the risks and achieve higher levels of expected utility. A study of the future migration intentions of Slovak returnees from the UK (Baláž *et al.* 2004b) pointed to important changes in the structure of individual migrants' networks after their first migration. More than half of the total sample had maintained contacts with friends and colleagues in the UK after returning, and these were considered potentially important for those considering permanent migration in the future. There were significant increases in their self-reliance to manage risks after the first move. Self-reliance was identified as being the most important way for reducing uncertainty in terms of obtaining a job by the returnees considering permanent migration. This was associated with increased self-confidence, as well as the experience of having lived abroad (Williams and Baláž 2005).

Competence and Decisions Under Uncertainty

We have already noted above that Fox and Tversky (1995) used experimental research methods to test Ellsberg's paradox relating to ambiguity aversion. Their experiments supported Ellsberg's assumptions about ambiguity aversion, but also suggested that ambiguity aversion only holds under comparative conditions where individuals compare outcomes from decisions under risk and uncertainty. Fox and Tversky repeated the Ellsberg paradox experiment, asking participants how much they would be willing to bet on the colour of a poker chip to be drawn from two bags in order to win a prize of $100: Bag A represented known risk (50 red poker chips and 50 black poker chips), and Bag B represented uncertainty (100 poker chips but the proportions of black and red are unknown). Half of the students performed a comparative task, considering the risk and uncertainty options together.

The other half performed non-comparative tasks, considering the two gambles separately.

It could be assumed that risk neutral, maximising individuals would be willing to bet up to $50 to win the prize, at which point the size of the gamble matches the statistical chance of winning. The results of the experiment indicated general risk aversion, as the average gamble in each game was lower than $50 (Table 3.1). The group in the comparative situation expressed strong ambiguity aversion and were ready to pay on average $24.34 for the clear bet (Bag A—known risks), but only $14.85 for the vague bet (Bag B—uncertainty). In contrast, the group in the non-comparative situation was ready to gamble similar sums under conditions of risk and uncertainty ($17.94 and $18.29). This resonates with migration. In some instances, individuals face choices between destinations about which there are varying degrees of uncertainty, and they can be expected to demonstrate ambiguity aversion. On other occasions, they have to make decisions about individual countries in isolation (perhaps responding to an invitation from a friend who lives there) and—according to the Fox and Tversky (1995) findings—there

Table 3.1   Testing the Ellsberg paradox: USA versus UK versus Slovakia, and migrants versus non-migrants

Amount willing to gamble (median values of bets)		
Fox and Tversky, Stanford University, risk and uncertainty bets, USD	*Risk conditions*	*Uncertainty conditions*
comparative conditions	24.34	14.85
non-comparative conditions	17.94	18.42
Baláž *et al.*, EU Bratislava, risk and uncertainty bets, EUR	*Risk conditions*	*Uncertainty conditions*
comparative conditions	22.70	18.29
non-comparative conditions	20.75	20.25
Williams and Baláž, UK survey, risk and uncertainty bets, GBP	*Risk conditions*	*Uncertainty conditions*
comparative conditions	20.66	13.76
non-comparative conditions	18.48	18.69
Williams and Baláž, UK migrants survey, risk and uncertainty bets, GBP, migrants	*Risk conditions*	*Uncertainty conditions*
comparative conditions	27.84	19.76
non-comparative conditions	25.39	25.59
Williams and Baláž, UK migrants survey, risk and uncertainty bets, GBP, non-migrants	*Risk conditions*	*Uncertainty conditions*
comparative conditions	19.25	12.57
non-comparative conditions	17.12	17.33

Sources: Fox and Tversky (1995); Baláž *et al.* (2011); Williams and Baláž (2013) and authors' own data

should be no significant difference in decision making in relation to whether they faced risk or uncertainty conditions.

The analysis can be extended by including the notion of competence. Not surprisingly, 'people prefer to bet in a context where they consider themselves knowledgeable or competent than in a context where they feel ignorant or uninformed' (Heath and Tversky 1991: 7). Fox and Tversky (1995) again used experimental methods to explore this: were students willing to bet more in hypothetical bets about average annual temperatures for a city they were familiar with than for an unfamiliar city? Again, there was evidence of ambiguity avoidance under comparative conditions, but this time in a context where competence was important: they were only significantly likely to be willing to bet more on the familiar city (risk) than the unfamiliar city (uncertainty) under comparative conditions. The authors of this book repeated variants of this experiment with a sample of 539 students at Bratislava Economics University (EUBA) in Slovakia, and this confirmed there was general risk aversion (Table 3.1). The importance of competence was also investigated. As these were economics students, the competence-related questions were changed from temperatures to inflation rates and language characteristics in Slovakia and Greece. This research confirmed the findings of Fox and Tversky: there were far greater differences in the preference for risk than uncertainty under comparative conditions. Although the differences under non-comparative conditions were smaller, they did not disappear completely. In a separate study, the authors found broadly similar results for a large sample in the UK, with willingness to bet being broadly similar when risky and uncertain conditions are considered separately, while strong ambiguity aversion was evident in a comparative situation.

Extending this research to migration, we can think of, for example, Slovak migrants having perceived greater competence in relation to the Czech language (a very similar Slavonic language), and how this would influence a decision whether to migrate to the Czech Republic or to Hungary, which has a very different language. Considered separately, potential migrants would be no more likely to choose the Czech Republic than Hungary, on the basis of perceived competence, but differential preferences would become much stronger in a comparative situation. Two subsamples of migrants and non-migrants within the UK data were also analysed (Table 3.1), and they demonstrated the same pattern of general risk aversion, and of ambiguity aversion under both separate and comparative conditions. However, we can also note that migrants were willing to bet more than non-migrants in all four possible gambles, suggesting they are more risk tolerant. Whether migrants were initially more likely to be risk tolerant (the selectiveness of migration), or whether their migration experiences (the transformative nature of migration) made them more risk tolerant, is a complex question of causality that cannot be answered with these data.

BOUNDED RATIONALITY, RISK AND MIGRATION

Haug (2008: 599) notes that 'rational choice theory includes different utility dimensions and takes into account different costs and returns. Unfortunately, the weighting of different utility factors, the transitive ordering and the connection between monetary and nonmonetary factors all remain under-specified'. The limits of neoclassical approaches in explaining migration decisions pointed to a need for more realistic models of human behaviour. Behavioural approaches to migration decisions have tried to consider a broader array of factors that influence decision making, including the difference between risk and uncertainty conditions, and the importance of comparative contexts. Lee (1966), who recognized that there are both positive ('pull') and negative ('push') factors, stressed the limits of computing utilities under the neoclassical approach:

> Needless to say, the factors that hold and attract or repel people are precisely understood neither by the social scientist nor the persons directly affected. Like Bentham's calculus of pleasure and pain, the calculus of +'s and –'s at origin and destination is always inexact.
>
> (Lee 1966: 50)

The emerging critiques of many of the assumptions of neoclassical models at that time were summed up by Herbert Simon:

> The rational person in neoclassical economies always reaches the decision that is objectively, or substantively, best in terms of the given utility function. The rational person of cognitive psychology goes about making his or her decisions in a way that is procedurally reasonable in the light of the available knowledge and means of computation.
>
> (Simon 1986: S210–S211)

Wolpert's (1965) paper on the behavioural aspects of migration was the first to consider bounded rationality and suboptimality in migration decisions. Wolpert's stress-threshold model assumed that potential migrants aspire to achieve a threshold level of utility. Decisions about whether or not to migrate are based on comparing this threshold to the utilities provided by particular migration destinations. Potential migrants have information on past rewards in their current place of residence, and try to extend these rewards to expected future rewards in a rational way. Incomplete information means that their knowledge of future rewards is necessarily subjective and rationality bounded, and that the outcomes of decisions are likely to be suboptimal. An important contribution of Wolpert's model was to separate migration into two stages: the decision to move and the actual behaviour. He also refined the concepts of action space, search behaviour, and place utility.

Some theoretical models of search behaviour and decision making have tried to formalise the effects of time constraints, beliefs, preferences and risk aversion in relation to the expected utility functions (Smith and Clark 1979). The stress-threshold models, and search-behavior and decision-making models, are more realistic than simple expected utility models which are based on assumptions of perfect information and rational decisions. They recognise that migration decisions are complex and include diverse pull-push factors. However, such models are very demanding in their data requirements and difficult to test. They have attracted more attention in residential mobility research than in migration studies: for example, Sun and Manson's (2010) study of housing search and locational choices in Minnesota.

The tension approach by Hoffmann-Nowotny (1981) tried to explain international migration within a general theory of societal systems, rooted in behaviourist concepts. It saw migration as a result of structural and anomic (i.e. psychological characteristics, or the emotional state of the individual) tensions encountered by an individual within a societal system. Strong behaviourist features can also be found in human capital theories, as presented by Chiswick (1978) and Bauer (1995), especially when they stress the psychological aspects of migration decisions. However, formalising a large number of pull-push factors is challenging and some psychological factors underlying migration decisions are difficult to measure in respect to expected utility. There is also a need to engage with notions of risk tolerance and aversion, and the behavioural economics literature that was emerging in this field.

RISK TOLERANCE AND RISK AVERSION

There is a considerable generic research on risk tolerance/aversion, associated with the work of Tversky and Kahneman (1974) and their associates, which mostly focusses on financial investment and health issues. Empirical research is largely based on experimental research methods, with individuals typically being asked to respond to hypothetical investment or gambling options. Only recently have researchers sought to extend the burgeoning body of work on risk tolerance/aversion influence to individual migration decisions and behaviour. Although this has an important contribution to make, there is a need to avoid over-simplistic, a priori theorisations about the relationship between risk aversion/tolerance and migration (Jaeger *et al.* 2007). It does not automatically follow that the more risk tolerant are more likely to migrate because of the risks and uncertainty that this entails. The relationship is necessarily contingent and, in some circumstances, there may be very high levels of risk and uncertainty about the future of jobs, income, security, etc., in the current place of residence, so that the more risk tolerant might be expected not to migrate.

The risk tolerance/aversion framework provides rich theoretical insights for migration studies. As Massey *et al.* (1993: 456) argued in their seminal review, it has remained difficult to express the probability of migration convincingly 'as a function of individual and household variables'. Socioeconomic characteristics may indicate which social groups have a higher propensity to migrate, but are unable to explain why specific individuals within these groups become migrants, while others do not. There are several possible reasons for this, including previous migration histories and social networks, as well as an array of sociopsychological characteristics. Risk aversion/tolerance provides one means of providing insights into the migration decision, but this remains relatively neglected, in part due to difficulties in operationalising these concepts.

MEASURING RISK TOLERANCE AND AVERSION

There is a long tradition of measuring risk attitudes in economics. Hundreds of studies on risk-aversion have been conducted since 1964, when John Pratt published his seminal work in *Econometrica* (see Table 3.2). The theoretical foundations of the Arrow-Pratt risk measure are quite straightforward, but measuring risk attitudes in practice is challenging, and the empirical findings have been surprisingly diverse. The diverse results and contradictions in measuring risk attitudes stem from: (i) the use of diverse types of reference variables (for example, changes in wealth, income, consumption); (ii) diverse definitions of the reference variables (for example, for wealth, should financial or total wealth be measured?); and (iii) diverse measurement methods (for example, objective versus subjective). The expected utility theory generally considers changes in utility in relation to changes in 'wealth', but is quite flexible about the reference variable: wealth, income, consumption and/or lottery bets are used to measure risk aversion.

For migrants, wealth is likely to be the most important variable—at least for labour migrants. This still begs the question of how wealth is defined. Individual wealth can be grouped into four major asset classes: (1) Financial assets (savings, insurance policies, pension claims); (2) Business assets (ownership or share in ownership of business); (3) Real estate assets; and (4) Human capital assets (net present value of expected lifetime income). Financial assets are considered less risky than business assets, human capital and real estate assets.

Financial assets are easiest to measure, followed by real estate and business assets, while the valuation of human capital is more challenging. Few studies have tried to compute this directly as part of the estimation of total wealth (Friend and Blume 1975; Schooley and Worden 1996; Halek and Eisenhauer 2001). Instead, most risk-aversion studies use financial wealth and/or income as the reference variable for computing the Arrow-Pratt coefficient. This approach is not unproblematic. The ownership of some risky

Table 3.2 Overview of selected large-scale risk-aversion surveys

Author	Data source	Sample size	Age	Male gender	Education	Entrepreneur	Higher income	Higher wealth	Asset structure	RRA coefficient	Reference variable	R^2	Estimate method
Objective risk attitude approach													
Halek & Eisenhauer (2001)	HRS (1992)	2376	–	+	0	–	n/a	$+^2$	F, R, HC	3.74	wealth	0.473	OLS
Pålsson (1996)	HINK (1985)	7904	–	+	0	0	0	0	F, R	10.8a (2.8b)	wealth	0.012	OLS
Schooley & Worden (1996)	SCF (1989)	2239	n/a	+	+	n/a	0	+	F, HC	1.24	wealth	0.480	OLS
Chang et al. (2004)	SCF (2001)	4442	–	+	+	+	+	0	F	n/a	wealth	n/a	logit
Subjective risk attitude approach													
Halek & Eisenhauer (2001)	HRS (1992)	7044	–	+	+	0	n/a	0	F	0.92	wealth	n/a	logit
Hartog et al. (2000)	Brabant Survey (1993)	2011	n/a	+	+	+	+	0	F	0.00154	lottery bet	0.053	OLS
Hartog et al. (2000)	Accountant Survey (1999)	1599	+	+	n/a	0	0	n/a	n/a	0.00077	lottery bet	0.024	OLS
Hartog et al. (2000)	GDP (1998)	17097	–	+	+	+	+	n/a	n/a	0.00034	lottery bet	0.040	OLS

(Continued)

Table 3.2 (Continued)

Author	Data source	Sample size	Age	Male gender	Education	Entrepreneur	Higher income	Higher wealth	Asset structure	RRA coefficient	Reference variable	R²	Estimate method
Barsky et al. (1997)	HRS (1992)	11707	–	+	+	+	+	+	F	4.00–8.00	cons.	n/a	OLS
Sahm (2007)	HRS (1992–2002)	10230	–	+	+	n/a	0	0	F, R	4.00–4.72	cons.	n/a	OLS
Yao et al. (2004)	SCF (1983–2001)	4442	–	+	+	+	+	+	F, R	n/a	n/a	n/a	logit
Sung & Hanna (1996)	SCF (1992)	2659	–	+	+	+	+	+	F, R	n/a	n/a	n/a	logit
Chang et al. (2004)	SCF (2001)	4442	–	+	+	+	+	+	F, R	n/a	n/a	n/a	OLS
Hallahan et al. (2004)	Pro Quest (1999–2002)	16461	–	+	+	n/a	+	+	F	n/a	n/a	0.238	OLS
Donkers et al. (2001)	CSS (1993)	3949	–	+	+	+	+	+	F	n/a	n/a	n/a	OLS
Dohmen et al. (2006)[3]	SOEP (2004)	21875	–	+	+	+	+	+	F, R	3.97–4.91	wealth	0.063	probit

Notes: + higher or increasing; – lower or decreasing; 0 = statistically insignificant; 1 = increases with square of wealth; 3 = general risk tolerance is the predictor of risk tolerance; n/a = variable or data not available; F = financial assets; R = real estate assets; HC = human capital assets; a = total wealth = F+R, b = financial assets only; cons. = consumption.

Source of comparison: Authors' review, based on the abovementioned references financial assets usually depends on education and/or income. Few low-income individuals have access to shares, mutual funds and/or financial derivates. Therefore, low shares of risky assets in the total financial assets of such individuals are misleading in terms of their risk aversion.

Human capital is likely to be the most important source of wealth for migrants, although some migrants also have significant business assets (entrepreneurial migrants), or real estate assets (often true of retired migrants). In part, this is age related. Young and skilled (middle class) migrants probably hold most of their wealth in human capital (and the lifetime stream of income expected from this). Migration, where it offers a higher return to human capital, is therefore likely to be particularly important to younger skilled migrants. In contrast, later in the life cycle, individuals are more likely to hold a larger share of their wealth in housing and financial capital, and as migration does not usually offer a higher return to these, they are less likely to migrate.

The distribution of wealth between low- and high-risk assets constitutes an 'objective' measure of risk aversion, and comes closer to the original Arrow-Pratt measure of risk aversion than do subjective measures. In real life, the structure of assets is difficult to estimate as data on wealth tend to be incomplete and/or confidential, and investment in housing, for example, is mediated by cultural expectations, habits and norms. Because of the difficulties involved in measuring risk 'objectively', most researchers rely instead on the measurement of 'declared' risk attitude, or subjective risk aversion, based on individual self-assessment either via questionnaires or interviews. Subjective risk attitude is also problematic, as it reveals attitudes rather than actual behaviour. The subjective risk attitude approach, however, has advantages: (i) relative ease of data collection; (ii) being less affected by external constraints and limits for reallocating risky assets; and (iii) helping to reveal psychological factors underlying risk attitudes.

Despite these reservations, the distribution of risk-aversion attitudes is remarkably similar across the diverse methods utilised, as evident by comparing measures based on lifetime income versus willingness to bet on a lottery. The Health and Retirement Study (HRS) in the USA contains a lifetime income question, whereby individuals are asked their attitudes to taking a new job with known risks of increased versus decreased income. This was broadly replicated by the authors in their own UK survey (Figure 3.1). The findings indicate that although both the USA and the UK samples were relatively risk averse, the UK sample was distinctly more averse in terms of the lifetime incomes associated with a job change: 83.9% in the UK, compared to 64.6% preferring the least risky option in the USA. Mobility seems to play an important role in willingness to take risks or, vice versa, willingness to take risks is an important precondition for mobility. In the UK sample, for example, the most risk-averse response was selected by 73.3% of the individuals with migration history, compared to 86.0% of those with no

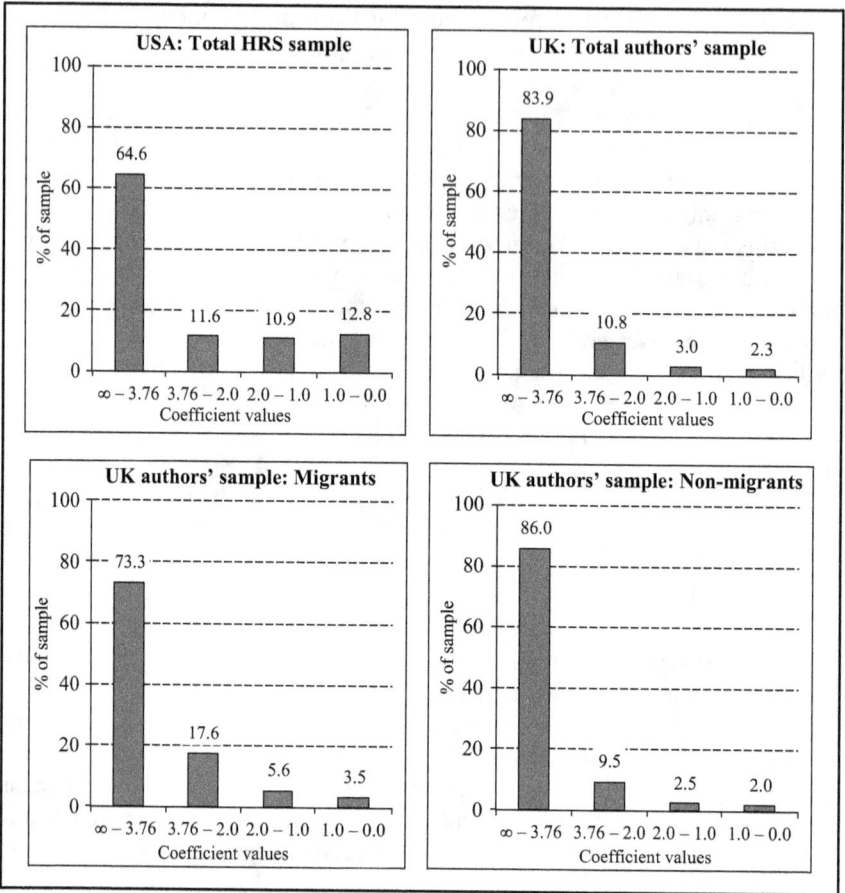

Figure 3.1   Percentage distribution of the Arrow-Pratt risk measure in the USA and UK populations, measuring the lifetime income gamble

Sources: Barsky *et al.* (1997) for the USA and authors' unpublished survey for the UK

migration history. The migrants also tended to be male, young and well educated—points that we return to later in this chapter.

The second measure, based on lottery questions, typically asks what individuals are willing to pay for a lottery ticket with a known prize and probability of winning (Hartog *et al.* 2000) (Figure 3.2). The shares of risk-averse individuals were broadly similar in the Brabant Survey (88.1%) in the Netherlands and the UK (79.8%) surveys. The most striking differences were to be observed in the risk seeking category: the proportion in the UK sample (10.0%) was six times greater than that in the Netherlands sample (1.6%). Nevertheless, it should be emphasised that the proportions of risk-seekers were relatively small in both populations, and the UK study used an online survey while the Dutch one was paper based: the

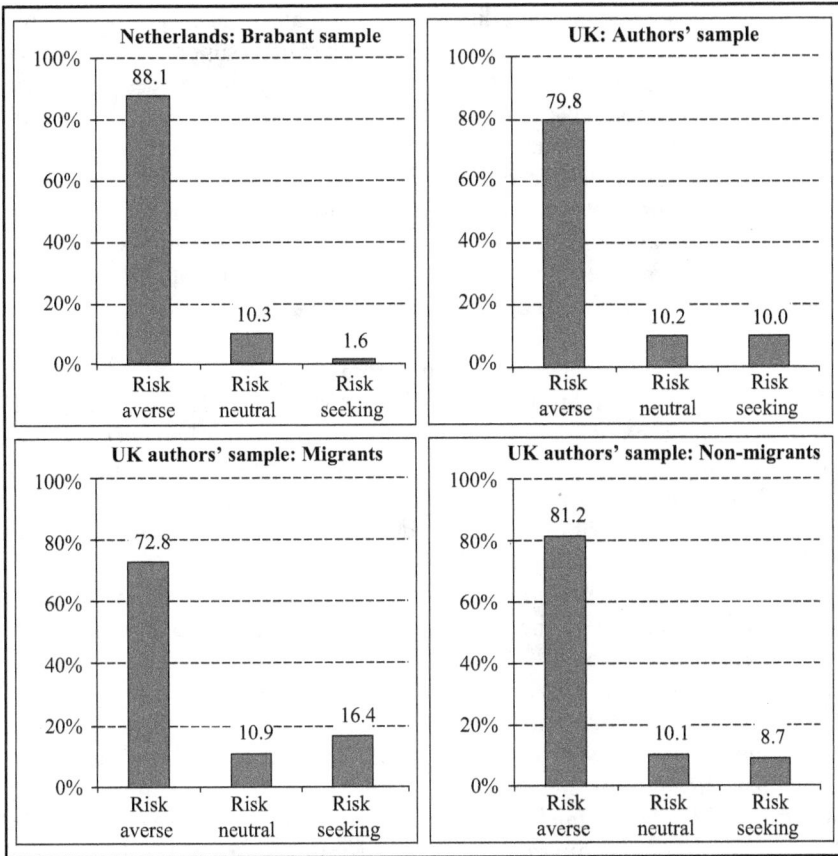

Figure 3.2   Percentage distribution of Arrow-Pratt risk measures in the Dutch and UK populations measured via hypothetical lottery gambles

Sources: Hartog *et al.* (2000) for the Netherlands and authors' own data for the UK

population in the UK study were more likely to be younger and better educated and, therefore, to be risk seeking. Again, there were notable differences in the proportions of migrants (72.9%) and non-migrants (81.2%) who were risk averse, and the proportion of migrants who could be considered risk seeking was twice as large as the equivalent proportion of non-migrants.

General Risk Trait

A persistent area of concern for behavioural economists has been whether there exists a 'general risk trait' which determines risky behaviour in all domains of life, such as drinking, driving too fast, or engaging in risky sports, or whether there are important differences in risk attitudes in specific domains such as migration The concept of a 'general risk trait' is supported by research in neurology and genetics, whereas social science research provides evidence of differences in risk tolerance across specific life domains. For example, an individual who engages in mountaineering may be averse to other forms of risk, such as migration. Recent research confirms that there probably is a 'general risk trait' based on biological and genetic factors, but that this is obscured by external factors and constraints. Moreover, dissimilar behaviours in specific domains of risk may be related to different levels of familiarity and perceived competence in these. Two major studies in the USA and Germany—based on large social surveys—have indicated the existence of a general risk trait.

Barsky *et al.* (1997) analysed data from the 1992 wave of the USA's Health and Retirement Study (HRS). The HRS sample (11,707 participants) were asked a range of questions on health, finances and behaviour in specific life domains (smoking, drinking, employment status, migration history). Barsky *et al.*'s study indicated statistically significant correlations between (subjective) risk-aversion and specific risky behaviours, such us drinking, smoking, migration history, and the absence of health and life insurance. The study also found 'tremendous variability in the behaviours, so only a small fraction of their variance is explained by risk tolerance (or any covariate)' (Barsky *et al.* 1997: 575): the estimated R^2 for risk tolerance and drinking, for example, was only 0.065, while that for the lack of health and life insurance was 0.073. However, and of particular relevance to this book, the highest R^2 (0.3) was reported for immigration history, indicating the particular relevance of this approach to migration studies. In another study based on HRS data, Halek and Eisenhauer (2001) found that those who have previously been migrants were significantly more risk tolerant than non-migrants in the USA (Halek and Eisenhauer 2001: 22).

The most comprehensive research on risk attitudes in Europe was undertaken in Germany, based on the longitudinal SOEP sample of some 20,000 individuals (Dohmen *et al.* 2005; Dohmen *et al.* 2006; Jaeger *et al.* 2007). The SOEP participants ranked both their general willingness to accept risk, as well as in specific domains: car driving, financial matters, sports and leisure, careers, and health. They identified relatively high correlations between general risk tolerance and the other five forms of risky behaviour (R^2 values varied, ranging from 0.474 to 0.609). However, the overall predictive power of willingness to take risks for explaining individual differences in risk taking were quite low in Dohmen's studies. As in the American studies, there appears to be a 'general risk trait', but expression of this trait may be subject to considerable constraints and external factors. Of particular interest is Jaeger *et al.* (2007: 3): analysing a different SOEP sample, they found that, after controlling for socioeconomic characteristics, willingness to take risks accounted for much of the residual variance in internal migration intentions. Nevertheless, there was still substantial unexplained variance, and this may be due to their reliance on general measures of risk tolerance, rather than migration-specific measures. In the Netherlands, van Dalen and Henkens (2012) provided evidence that sensation seeking (rather than risk seeking) was significant in explaining migration intentions, but their broad model did not analyse risk tolerance per se.

More recently, Williams and Baláž (2013) analysed migration and risk tolerance in the UK, drawing on a large sample of 4,528 individuals. Table 3.3 summarises differences in attitudes to risk and competence to

Table 3.3 Tolerance of risk in the UK: migrants versus non-migrants

	Non-Migrants	Migrants
1: Please tell us about your attitude to risk when investing money.	1.69	2.05
2a: I often drive too fast.	3.74	4.25
2b: I often smoke too much.	2.67	3.22
2c: I often do risky sports.	2.43	3.48
2d: I often drink too much.	3.54	4.10
3: Suppose that you are the only income earner in the family, and you have a good job guaranteed to provide your current (family) income for your lifetime. You are given the opportunity to take a new and equally good job, but with a 50-50 chance it will double your (family) income and a 50-50 chance that it will cut your (family) income by a third. Would you take the new job?	1.77	2.09
4: What do your best friends think about your attitude to risk?	2.58	2.24
5: Flipping a coin, 50-50 chance to win or lose, pay GBP 1,000 if you lose.	93.32	93.55
6: Do you know many people living in, or who have returned from, the country/countries you considered moving to?	3.99	5.07
7a: How much would you bet on a green ball? Ellsberg Q1	19.25	27.84
7b: How much would you bet on a yellow ball? Ellsberg Q2	12.57	19.76
7c: I am ready to bet. Ellsberg Q3b	17.12	25.39
7d: I am ready to bet. Ellsberg Q4b	17.33	25.59
8a: I adapt more flexibly to new situations than my friends.	5.85	6.55
8b: I manage my problems better than my friends.	6.05	6.11
8c: I am able solve problems related to travelling abroad better than my friends.	5.60	6.79
8d: I would adapt better to living abroad than my friends.	5.45	7.01
8e: I am willing to take more risk than my friends.	4.77	5.91

Source: Author's own survey; see Williams and Baláž (2013)

manage risk across a number of domains. Migrants consistently had greater propensity to take risks in terms of every single measure recorded: investment, driving, smoking, risky sports, smoking too much, job changes with more risky earnings, flipping a coin, and betting on green versus yellow balls in the Ellsberg paradox (that is, on risk versus uncertainty), under both comparative and non-comparative conditions. The last two measures relate to a pure chance environment, while the other measures are those in which some degree of competence can be exerted by the individual. Migrants are generally more likely to be willing to take risks than non-migrants. Migrants are also more likely to have perceived competence to manage risks, whether in respect of their general competence to manage change (flexibility and problem solving) or specifically in relation to travelling and living abroad. They also consider that they are more willing to take risks than their friends, but interestingly they consider that their friends are less likely to consider them as willing to take risks, than do the friends of non-migrants. They generally have more contacts who live abroad, or have lived abroad, which is an important resource in managing risk, and one which probably contributes to their perceived greater competence in managing risks.

Finally Box 3.2 provides a summary of the relationships between gender, migration and risk in Slovakia, indicating significant differences between men and women, and women migrants and women non-migrants.

Box 3.2   Gender, Migration and Risk: A Study of Slovak Students

The authors studied the attitudes to risk of a sample (n = 359) of Slovak students, a group who are relatively socioeconomically homogenous. This also included a substantial subsample who have had temporary international migration experiences (longer than three months), mostly on short-term work or study placements. Their attitudes to risk were assessed under experimental conditions, which measured their willingness to take risks on hypothetical gambles under different conditions.

In both competence-informed and pure-chance contexts, the study found evidence of ambiguity aversion (see Table 3.1). This applied to both the non-comparative conditions indicated by Ellsberg (2001), as well as under comparative conditions, which Fox and Tversky (1995) considered were critical in differentiating willingness to bet on risk versus uncertainty.

The study also considered differences in risk taking between men and women, and between migrants and non-migrants. It confirmed, as has been demonstrated in other studies, that there are important gender differences when making decisions under both risk and uncertainty conditions, with men being more willing to gamble than women. Additionally, and for the first time, it analysed differences between migrants and non-migrants in respect to risk versus uncertainty. Inevitably, this was inter-related with gender. Women migrants were more willing to take risks than women non-migrants, in a pure-chance risk context, as opposed to a competence environment. As expected, men were more likely than women to take risk in a competence environment, reflecting overoptimism and overconfidence.

The research did not find any significant differences between men and women in terms of their self-assessed capabilities and willingness to take risk. Perhaps the most striking finding, however, was that women had significantly higher risk tolerance than women non-migrants, irrespective of their self-assessed capabilities. In contrast, there were no differences between migrant and non-migrant men in this respect. This may reflect the fact that the types of migration involved—mostly short-term work or study periods—are relatively low risk, and therefore risk tolerance was not a significant differentiator of who did and did not migrate amongst men. In contrast, amongst women—who generally had lower levels of risk tolerance than men—this did seem to be an important differentiator of who had previously migrated.

Sources: Fox and Tversky (1995); Baláž *et al.* (2011); Williams and Baláž (2013) and authors' own data.

THE SOCIO-DEMOGRAPHY OF RISK ATTITUDES

Behavioural research on risk aversion suggests there are significant differences in willingness to take risks between socioeconomic and sociodemographic groups, and amongst individuals within these groups (Halek and Eisenhauer 2001; Hartog *et al.* 2000; Barsky *et al.* 1997; Sahm 2007; Yao *et al.* 2004; Sung and Hanna 1996; Chang *et al.* 2004; Hallahan *et al.* 2004; Donkers *et al.* 2001; Dohmen *et al.* 2006).

Men tend to be more risk tolerant than women. This is reported in most large-scale studies (Barsky *et al.* 1997; Halek and Eisenhauer 2001; Hallahan *et al.* 2004, Pålson 1996) and the findings tend to be highly statistically significant. Barsky *et al.* (1997: 550) noted that the biggest difference was in males' greater propensity to choose the highest risk option when considering potential job changes that involved lifetime earnings risks. However, the finding about males being more risk seeking may be context-specific. Men account for most household heads in many surveys, including the HRS, and it is not clear how far they express personal as opposed to household opinions, which are necessarily difficult to differentiate. There is a link here to the new economics of migration and households risk diversification (Chapter Five). Surveys based on lottery questions (Hartog *et al.* 2000; Donkers *et al.* 2001) also found higher risk tolerance among men. A meta-analysis of 150 studies (Byrnes *et al.* 1999), for example, found that men were more risk tolerant in 14 out of 16 observed types of risk behaviour. However, a range of factors influences risk-taking attitudes in different contexts. There is some evidence (Daruvala 2007; Ronay and Kim 2006) that higher risk aversion by women may be a socially facilitated phenomenon. Most research on risk taking focusses on financial decisions, reflecting the prominence of the topic and the vast amounts of data produced by financial markets and institutions. Studies of financial risk taking, however, are made in a competence-informed context, where prior knowledge is brought into play. Men generally perceive themselves to be more competent in financial affairs and are ready to gamble higher sums than women. This poses the question of whether migration is an arena of pure chance or competence-informed.

Younger individuals are less risk averse than older ones. Barsky *et al.* (1997) found the youngest (under 50) and the oldest (70+) cohorts were most risk tolerant, while middle-aged cohorts were more risk averse. The HRS-based studies (see also Sahm 2007; Halek and Eisenhauer 2001), however, overrepresent participants aged 50+. Lottery-based questions generally indicate increasing risk aversion with age: the only exception is the rather unusual sample analysed by Hartog *et al.* 2000. Dohmen *et al.* (2006: 26) note that 'Age decreases the probability that an individual is willing to take risks in all five domains, but has a particularly large impact in the domain of sports and leisure, and a relatively small impact in financial matters'. These findings indicate that increasing (perceived) competence may offset decreases in 'pure risk' attitudes with increasing age. Some surveys on financial behaviour (Sung and Hanna 1996; Hallahan *et al.* 2004), however, indicate that risk tolerance declines at an increasing rate with increasing age. Sahm (2007: 24) found that while aging does affect risk-tolerance levels, other changes in individual circumstances, including the loss of a job or the end of a marriage, do not. Therefore, decreasing risk tolerance over time probably is not related to major (negative) life events, such as illness or divorce or loss of jobs, but to aging per se, which can be understood as physiological changes including hormonal and brain chemistry changes.

Educated individuals tend to be more risk tolerant. Barsky *et al.* (1997) found that individuals with less than 12 years and over 16 years of education were the most risk tolerant. However, Halek and Eisenhauer (2001), Hartog *et al.* (2000, GDP data) and Hallahan *et al.* (2004) found statistically significant increases in risk tolerance with increasing education on all levels. Evidence from surveys based on both objective and subjective risk attitudes indicate that risk aversion generally decreases with increasing education attainment as a result of the competence effect.

Individuals with higher income and wealth are less risk averse. Barsky *et al.* (1997) found that families with the lowest and the highest income and wealth levels had the highest risk tolerance. This was consistent with their findings for education, which is not surprising given that education is an important predictor of income and wealth. Halek and Eisenhauer (2001) found increasing risk aversion up to a certain level, at which point the risk aversion turned to risk seeking. The effects of income on risk aversion are obscured by different definitions of wealth (discussed earlier) and entangled with the effects of education and age, as well as different vulnerabilities to the effects of income losses (Carretero 2008). Adams (2001: 66) contends that 'poverty will affect the perception of rewards and dangers and can induce people to take extra risks. There is a steep social-economic class gradient to be found in accident rates, with the poorest experiencing much higher rates than the wealthiest.' This is supported by research on the attitudes of Mexican migrants to the USA towards environmental risks (Vaughan and Dunton 2007): those who have limited employment choices tend to minimise scientific risk information when making risk judgements.

The self-employed are more risk tolerant than employees. This finding is indicated by most large-scale studies (Barsky *et al.* 1997; Hartog *et al.* 2000; Donkers *et al.* 2001; Dohmen *et al.* 2006; Sung and Hanna 1996). Halek and Eisenhauer (2001) surprisingly found that the self-employed tend to be overinsured (contrary to much of the literature on entrepreneurship), but explained this by their limited access to other forms of insurance, such as unemployment insurance. The relationship between risk and self-employment remains an intriguing one because migrant self-employment and entrepreneurship can be understood in terms of a response to the existence of glass ceilings in the labour force for migrant employees (OECD 2010), as well as to their greater willingness to take risks.

There are national differences in risk aversion. The willingness to take risks varies across countries, even amongst the more developed economies. For example, Fehr *et al.* (2006) provide evidence that Germans are less willing than Americans to take risks. And if risk-averse individuals migrate, they are more likely to move to countries with stronger rather than weaker welfare systems, in order to reduce the risks associated with unemployment (Heitmueller 2005).

The above review of how socioeconomic factors impact risk tolerance was also confirmed in the authors' own surveys of attitudes to financial risk in the UK and Slovakia (Figure 3.3). Risk tolerance decreased with age, increased with education and was higher for men than for women:

- *Age:* Above-average and substantial financial risks were acceptable for 29.5% of the UK respondents aged 18–25, but for only 14.1% of respondents aged 65+. In Slovakia the respective shares were 18.4% of respondents aged up to 29 and 7.7% of respondents aged 60+.
- *Gender:* Above-average and substantial financial risks were acceptable for 29.8% of UK male respondents, but only for 16.6% of female respondents. In Slovakia the respective shares were 20.3% and 6.3%.
- *Education:* Above-average and substantial financial risks were acceptable for 14.8% of UK respondents with primary and lower middle education, for 24.6% of respondents with upper middle education, and 30.1% with tertiary education. In Slovakia the respective shares were 5.4%, 10.3% and 18.7%.

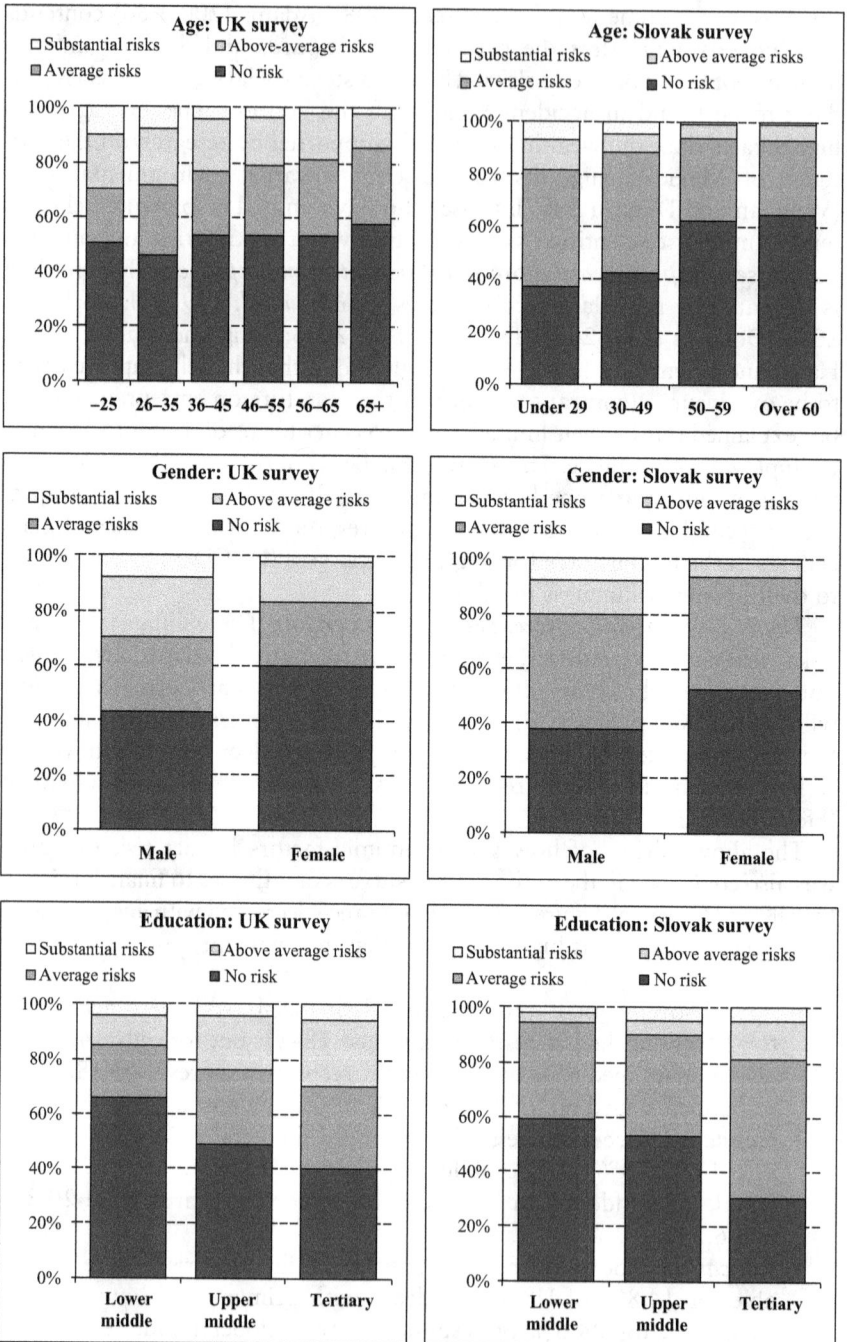

Figure 3.3 Socioeconomic correlates of financial risk tolerance

Sources: Authors' surveys in the UK (N = 4,528) and Slovakia (N = 731): see Williams and Baláž (2013).

Correlation coefficients measuring relations between risk attitudes and exogenous socioeconomic and sociodemographic variables indicate the strength of statistical associations but do not provide evidence of causality. Halek and Eisenhauer (2001) considered that the a priori evidence for the direction of causality is least clear in the cases of education, marital status, employment status and income/wealth. Here we consider the example of education. Education may have provided competence to manage risks. It is also true, however, that perceived knowledge in one specific domain which can be extended to other domains can result in overconfidence and risk seeking. Just because an individual has a relatively higher education level does not necessarily make him/her better at managing migration risks. The direction of causality is also not clear. Higher risk tolerance may have informed the decision to participate in secondary and higher education (and leave or stay in school). Education therefore can be an outcome of self-selection related to risk attitudes. Similar issues relate to the causality between migration and risk tolerance.

Another issue concerns how risk aversion/tolerance changes over the course of the migration cycle. Do migration experiences modify willingness to take risk, perhaps as a result of enhanced risk resilience or through having created social networks which reduce risk levels for individuals in the external environment (Brunnermeier and Nagel 2005)? Arguably, successful migration experiences provide learning experiences, and the acquired competences make migrants more risk tolerant when considering further migration. However, Jaeger *et al.*'s (2007: 13) findings appear to contradict this: they undertook ex ante and ex post regressions in relation to when the risk question was posed in their panel data, while controlling for previous migration in the post ante analysis. The results indicate stability in risk tolerance over time: it does not seem to be modified by the migration experience. Although there is a lack of other empirical evidence to confirm Jaeger *et al.*'s findings for migration, we can draw on Sahm's (2007) work on attitudes to risk in employment. Although there was a moderate decline of risk tolerance with age, and changes over the business cycle, the main finding was a well-defined and relatively stable set of risk preferences, which accounted for 80% of the variation in their data. Neither changes in wealth and income, nor personal events, have significant impacts on the willingness to take risk; instead, there is constant relative risk aversion over time. 'Attitudes toward risk may move over the business cycle, but there is no evidence this translates into a permanent shift in risk tolerance' (Sahm 2007: 25).

Risk Seeking

Risk seeking describes an individual whose utility function's second derivative is positive. Such an individual would willingly pay a premium to assume particular risks. Initially, this seems a surprising concept, at least in the context of migration studies. Most people are risk averse and in experimental research even find 'fair games' (offering equal probabilities for the same amount of potential good and bad outcomes) unattractive. If they were to enter a risky game, they would demand either a higher probability of winning (more than 50%) or a higher payoff for the positive outcome than for the negative one. The difference between the expected utility provided by a 'sure win' and the expected utility provided by the risky bet is the *risk premium*. In terms of wealth, the risk premium is the difference between the *certainty equivalent* and expected wealth. The certainty equivalent is the guaranteed amount of wealth that a decision-maker considers equal to a risky asset. The risk premium is an amount of money that the individual is ready to pay for avoiding risk. For risk-averse persons (i.e. the majority of the population), the risk premium is positive. In contrast, the risk premium is negative for risk-seeking individuals who are willing to pay for the pleasure or utility of participating in the risky endeavour.

How do these ideas apply to migration—are there risk-seeking migrants? Hamilton (1978) applied the term 'adventurer' to some of the migrants who were attracted by the various nineteenth century gold rushes He specifically recognized adventurism as a form of 'risk behaviour', and that 'to undertake such risks requires audacity—or foolhardiness or courage depending on one's perspective—that is not found in every individual, or even every stratum of society' (Hamilton 1978: 1467). However, he did not engage with the concept of risk seeking as such, let alone its understanding in terms of utility, instead considering that adventurism was driven by economic, political and social gains.

Risk seeking is more likely to be found in context of shorter term migrations undertaken by many young adults, often as part of a gap year, or some forms of backpacking (Elsrud 2001). Participants in such temporary migrations have various motivations, ranging from travel and tourism to adventure seeking. Green *et al.* (2008), for example, found that adventure seeking was an important motivation for 34% of New Zealanders migrating to Australia. At least a small minority are willing to pay (in the form of travel and other costs) to engage with risk. These are the risk seekers who gain utility from confronting and managing such risks, which also provide them with self-confidence and peer esteem. Giddens (1991) echoes this view in his discussion of adventurous travelling, which he understood as the 'active courting of risk' (Giddens 1991: 124), in a society preoccupied with the uncertainty of the future. This resonates with Lyng's (2008) notion of 'edgework', which recognizes 'the seductive power of the risk experience' (p. 120; see Chapter Six).

CONCLUSIONS: IN SEARCH OF ALTERNATIVE EXPLANATORY ECONOMIC FRAMEWORKS

No theory can explain all aspects of behaviour, whether migration or any other form of behaviour, because it is immensely complex and diverse. Although in many ways an advance on neoclassical theories, Tversky and Kahneman (1992) were able to explain just about one half of the total variation in the data produced in their laboratory experiments, and this was considered to be an exceptionally high proportion in behavioural studies. In part this reflects the complexity of decision making. Researchers have therefore sought to further advance the understanding of economic decision making by developing new perspectives to engage with complexity and incomplete knowledge. Much of this work remains in the early stages of development. There are large numbers of competing decision-making theories in this field, and here we consider two of the most frequently discussed models: (a) probabilistic and (b) nonprobabilistic. Both address how individuals seek to reduce complexity in decision making.

Probabilistic Theories

Rank-dependent utility models (Quiggin 1982; Tversky and Kahneman 1992) assume that choice among alternatives depends not only on objective probabilities and the size of the outcomes, but also on the ranking of the outcomes. Extreme outcomes are considered to attract most attention in the consideration and the decision-making stages, despite their low probabilities. For example, a migrant presented with a number of job opportunities is likely pay greater attention to the highest and lowest potential earnings. This model reinforces the view that the weighting employed in decision making is not linear, and it also implies there is a need to take into account the complexities of how risks and uncertainties are (implicitly) taken into account. There is a family of rank-dependent utility models that can be derived from this proposition, including the cumulative prospect theory discussed in Chapter Two. Of particular note are the transfer of attention exchange models (TAX) (Birnbaum and LaCroix 2008), which are based on the idea of configural weights. The configural weight models recognise the same psychological phenomena as the expected utility and prospect theories, including risk aversion. They also acknowledge the nonlinear transformation of probabilities: that is, the importance of decision weights. In particular, the TAX models recognize the importance of how the alternatives are configured in *each* decision-making event: the same alternative will be assessed in relation to the available alternatives, and these vary from situation to situation. If the decision-maker is risk averse, the alternatives with higher probabilities attract more attention than those which have lower probabilities. The TAX models try to capture the variability of human decisions. They introduce new assumptions and parameters to models of decision making and provide a better description of the analysed data than, say, prospect theory.

For example, in terms of migration, the TAX models would contend that how a potential migrant from the Ukraine evaluates Spain as a destination against the USA depends on what other countries are included in the comparison—whether, for example, it is Bulgaria or Hungary. The risk (probability) of obtaining a job at a particular salary level (an outcome) in Spain is the same in both decisions. However, the alternative outcomes and probabilities of the comparator countries are different. In the first example, the attention of the decision-maker is transferred to the high probability of getting a job in the USA, where unemployment rates are low compared to Spain, Hungary and Bulgaria. Although wages are higher in Spain, the comparative framework makes it more likely that the migrant will select Hungary. In the second example, the attention of the individual is drawn to Bulgaria because of the low probabilities (high unemployment) and low outcomes (low wages). This increases the likelihood of the migrant choosing Spain, because of the comparison with its higher outcome. These migration examples remain purely hypothetical as there has been no attempt to apply such heuristics to migration to date.

Nonprobabilistic Theories

Probabilistic decision models were developed and tested in controlled laboratory environments. Experiments are performed via rather abstract tasks which seek to simulate real-world situations, but where the values of outcomes and their respective probabilities are mostly known to the participants. Real life, of course, is complex. Most individuals do not know either the exact value of the potential outcomes or their probabilities. The potential migrant does not know the exact income he/she will obtain or the probability of achieving this. Even where friends or family are arranging the job, there is still some uncertainty.

Some relatively new theories have sought to engage with the issues of strong assumptions about known outcomes and probabilities. The assumption of known probabilities is particularly problematic, as an overwhelming majority of future events have uncertain distributions of outcomes. A prospective Indian IT specialist migrating to the UK, for example, may have his/her beliefs about the availability of jobs and salaries, but the real chances of getting a particular job/salary is difficult to compute. Alternative theories to probabilistic theories address these issues by assuming that decision-makers use fuzzy logic or the info-gap theory which deals with extreme uncertainties.

Whereas probabilistic logic considers the 'probability' or 'likelihood' associated with certain events, *fuzzy logic* considers degrees of truth (Zadeh 1965). Fuzzy logic considers how much of a variable is within a set. For example, imagine that a newly arrived migrant considers that the likely annual income in a destination country is £5,000–30,000. The migrant is considering two concepts about his/her income: 'poor' income is anything below £5,000 and a 'good' income is anything better than £30,000 pounds. In these terms, an income of £20,000 pounds is 60% 'good' and 40% 'bad,' that is, in relation to £30,000. The proportions of 'bad' and 'good' income (or membership of these groups) determine the migrant's decisions. Therefore, it could be argued that if his/her income is lower than £5,000, the individual will return to the country of origin or migrate to a different country. With an income of £5,000–30,000, the individual stays in the destination country. If his/her income is greater than £30,000, the migrant might, for example, invite his/her family to join them. The concept of good/bad is subjective and depends on the individual's perspective. Thus, another prospective migrant may set different reference values for 'good' and 'bad', say £10,000 and £40,000. With an income of £20,000, he/she would consider that only 50% of this income was good, but the decision about staying would be the same as for the first migrant with the same income. Fuzzy logic has not yet been applied to migration studies, but it has potential to provide insights into decision making.

An alternative approach, the *info-gap decision theory*, analyses judgements and decisions taken under severe uncertainty (Schwartz *et al.* 2011). Unlike probabilistic theories, the info-gap decision theory does not rely on probabilistic distributions, but considers the deviation of errors. This is the difference between the real value of a parameter and its estimate, that is, it is an estimate of the deviation of your 'guess' compared to reality. Decision making involves three phases: (i) uncertainty, (ii) robustness/opportuneness, and (iii) decision making. The decision-maker first estimates the unknown value of the parameter and then examines the distance between this and other values of the parameter. Then the robustness of the model is examined, that is, how acceptable a likely deviation from reality, and its consequences, are for the decision-maker. A minimal and the desired values of the outcome are set and the uncertainty associated with each decision is estimated in relation to these. The next step examines the robustness and opportuneness of potential decisions and an optimal solution is sought.

Next we consider how this could apply to a prospective migrant. Income is the key parameter but it is uncertain. An income of £20,000 per year provides a decent living standard, an income of £12,000 means living on the edge, and an income of £30,000 pounds could enable the migrant to bring his/her family to the destination. The estimate of the unknown parameter is £20,000. The robustness function is £8,000 (£20,000 – 12,000), and the opportuneness function is £10,000 (30,000 – 20,000). If the migrant erred in his/her estimate by more than £8,000, then serious problems of survival would arrive. If he/she is certain about earning an income of £12,000–20,000, there is neither a serious problem nor an opportunity to bring the family to the destination. Now imagine that the migrant considers an alternative migration destination. The expected annual income is lower (£16,000 pounds), but minimal living costs also are also lower (£9,000 pounds) and the opportunity to bring your family requires an income of £28,000 pounds. The robustness function is £7,000, and the opportuneness function is £12,000. If the migrant makes the same errors for both sets of estimates, the second destination is both less robust and less opportune. Therefore, the migration decision will be to go to the first country. Individual migrants are unlikely to follow this exact course of decision making, but the approach is illustrative of how they reconcile a set of goals (in this case, survival and bringing family members) with uncertainty in relation to future income at the destination. This approach has not yet been applied to migration studies, but Wolpert's (1965) paper foresees some issues in decision making under severe uncertainty.

New Territories for Migration Research?

Since the 1990s behavioural economics has expanded into new territories and developed strong interactions with other fields of social science (anthropology in particular; see Chen 1999), neural sciences, behavioural genetics (Cesarini *et al.* 2009; Zyphur *et al.* 2009), biology (Sacredote 2002) and evolutionary biology (Eisenberg *et al.* 2008; Matthews and Butler 2011). Mainstream economic theory has so far avoided these fields, as they are difficult to integrate into general microeconomics theories. Here we summarise some findings from these more interdisciplinary studies, which relate to risk-taking and migration research, and which offer interesting future directions for migration researchers.

Some behavioural genetic studies see migration behaviour as a kind of evolutionary adaptation to a changing environment. Studies on genetic influences have explored links between dopamine-related genes, personality traits and behaviours, and reward-related brain activation. They focused on the 7R and 2R alleles (particular genetic variations) of dopamine receptor genes (DRD4):

- Chen *et al.* (2004) were the first to suggest that human migration may have been speeded up by mutation (alleles) of the DRD4 gene linked to risk-seeking behaviour. Individuals with such behaviour, in theory, were better prepared to cope with dangerous situations and unstable environments, and, in turn, produced more offspring than individuals with more sedentary traits. Chen *et al.* compiled data on 2,320 individuals coming from 39 different populations. They found that population groups with more intensive migratory patterns, compared to sedentary populations, showed a higher proportion of long alleles for DRD4. Chen *et al.* (2004: 320) argued for the evolutionary origin of differences in the DRD4 distribution in migratory and sedentary populations.
- There is some evidence that specific alleles of the DRD4 gene may be more useful for migrant populations than for sedentary ones. Eisenberg *et al.* (2008) surveyed the population of adult men of the Ariaal tribe in northern Kenya. The Ariaal used to be traditionally nomadic pastoralists, but some of them settled in the 1970s. Eisenberg *et al.* compared body mass index (BMI), fat free mass (FFM), arm muscle area plus bone area (AMPBA) and height with the frequency of the 7R allele and found that the '7R allele had a positive association with the BMI, FFM and AMPBA among nomads, but a negative one among settled

individuals' (Eisenberg *et al.* 2008: 173). The 7R allele is related to greater food and drug cravings and novelty seeking—indeed, it is often labelled the novelty gene. It may increase the ability of young nomads to defend livestock against raiders or to locate food and water sources. Such predispositions might be relatively less beneficial for settled communities practicing agriculture, selling goods on markets or attending school. Eisenberg *et al.*'s research suggested that genetically induced differences in risk seeking may be more or less effectively expressed in diverse environments (over time the allele would probably disappear in the sedentary population, and increase in incidence in the nomadic population—being a burden to the first group, helpful to the second).

- Mathews and Buttler (2011) looked at the frequency of two DRD4 alleles (7R and 2R) in 18 indigenous populations spread along the routes humans took from Africa to Europe, Asia and the Americas. They established that the distribution of the 7R and 2R alleles was not random, and there was a statistically significant link between the alleles frequency and migration distance travelled: the greater the distance, the higher the incidence of the particular alleles. Mathews and Butler (2011: 388) assume that mutation of the DRD4 gene is a relatively recent event (40,000–50,000 BP) and accounts for positive Darwinian selection. It roughly coincides with the first observed wave of out-of-Africa migration, and increased frequencies of 2R and 7R of the DRD4 gene have been effectively selected by repeated generations of migrants.

The mainstream theories of migration may yet find a way to integrate findings from anthropology and evolutionary genetics, but risk-seeking behaviour is a complex phenomenon. Except for the hereditary components, risk seeking is determined by the abundance of biological, psychological and/or economic and social factors. The main value of the behavioural genetics for migration research is that it may help explain those components of migration behaviour, which cannot be explained by other factors. If migrants are self-selected in terms of their higher tolerance of risks, for example, some part of the tolerance may refer to higher occurrence of particular genetic variations.

4 Risk and Complexity
Decision Making Under Different Information Conditions

INTRODUCTION

The previous chapter considered how behavioural economists have sought to address what were seen as the limitations of neoclassical models. They recognised that decision making occurs within a framework of bounded rationality, rather than optimising perfectly informed decision making. In particular, we considered three main themes that have considerable resonance for migration studies: the difference between risk and uncertainty, the importance of risk tolerance/aversion/seeking, and the importance of perceived competence to manage risk. These approaches bring a more nuanced approach to economic analyses of risk in migration studies, but there are still limitations in the extent to which they can explain how individuals respond to imperfect information, and the sheer complexity of the issues involved in decision making.

The importance of incomplete information has long been recognised. Risk attitudes imply that people are more likely to migrate to locations about which they have good quality information, available at a reasonable cost. This assumption may help explain why many migration flows are between neighbouring countries, or among countries having strong cultural, historical and language ties. Individuals with high levels of risk tolerance, on the other hand, are more likely to engage in long-haul migration to destinations about which they have limited information.

Da Vanzo (1983) demonstrates that perfect information and perfect foresight were prerequisites for the potential migrant to be able to assess effectively the advantages and disadvantages of migrating to a different place. Tunali (2000) extends these arguments to return migration, emphasising that such moves can only be understood if the role of uncertainty is acknowledged. Applying these ideas to return migration from Sweden to Finland, Saarela and Rooth (2012: 1893) have argued that one reason for return migration is because of 'mistakes generated by uncertainty in the initial migration decision. Imperfect information about the economic conditions faced at the destination leads to the decision to return migrate'.

This chapter considers the complex decision-making models that have been developed to provide insights into how individuals seek to deal with both incomplete, and too much, information. This approach remains grounded in the notion of bounded rationality, and therefore shares some of the reductionist features of the theories discussed in Chapters Two, Three and Four. To the best of our knowledge, complex decision-making models have only been applied to migration studies to date in our own studies, but the approach is discussed here because of its potential to advance understanding of how migrants deal with risk and uncertainty. The approach uses experimental research methods to simulate real decision-making conditions.

MIGRATION DECISIONS: UTILITY, RISK AND COMPLEX CHOICES

The complex decision-making approach originated in consumer decision making, and while we need to be cautious about how this is applied to migration—which is an investment and consumption decision, with very substantial life-course implications—it provides useful insights. Traditional models of consumer decision making depicted complex choice as 'multi-staged' and complex. Several factors are considered to trigger problem recognition by individuals before initiating a sequence of actions and decisions, leading to an outcome of 'satisfaction' or 'dissatisfaction'. This involved a 'traditional five step classification, i.e. the cognitive decision sequence of problem recognition/pre-search stage, information search, alternative evaluation, choice, outcome evaluation' (Erasmus *et al.* 2001: 83). These models are based on a rationalist approach to decision making and assume extensive and detailed weighing and evaluation of the attributes of different objects or the utilisation of different options (perhaps different destinations for migrants) in order to arrive at a satisfactory decision (Solomon *et al.* 2006: 268). However, rational goals were still assumed to be expressed in economic or objective criteria such as price, size and/or capacity.

Utility maximisation is at the heart of traditional economic theory. It is assumed that all goals (say obtaining food and shelter, establishing a family, achieving social status) can be defined in terms of a single common denominator, utility, whether in terms of maximal pleasure or minimal pain. If there is a common denominator (utility) for all human activities, then the final goal of every decision-making event is to maximise utility from all activities. Individuals should prefer those goals, or combination of goals, which result in the maximum possible utility. Pursuing the goal of maximal utility assumes that individuals would act selfishly and rationally. Utility maximisation is also inherent to prospect theory (Kahneman and Tversky 1979; Tversky and Kahneman 1992), considered earlier. Prospect theory computes utilities differently from expected utility theory (von Neumann and Morgenstern 1944), but nevertheless assumes that individuals want to achieve maximal utility (prospect). Most economic theories also assume that individuals know in advance what their goals are and how to achieve these. They are also assumed to know and accept the distribution of risks attached to the potential outcomes of their decisions: otherwise they could not act rationally to maximise utilities.

Psychological sciences take a different view of goal setting. They assume that: (a) individuals may simultaneously pursue multiple goals, and (b) these goals may be conflicting rather than complementary (Kranz and Kunreuther 2007). Moreover, goals need not be known/set in advance, but are constructed in the course of decision making. Plans are instruments for mobilising resources so as to achieve goals, and a plan may mobilise resources for simultaneously achieving multiple goals. A migrant, for example, may want a higher wage and therefore signs a contract to work on an offshore oil field. The same migrant, however, may also want to be more satisfied with his/her life. Working on a remote drilling platform presents limited social opportunities, but offers high wages. In the face of multiple (and potentially conflicting) goals, the different goals compete for the attention of the decision-maker. Giving selective attention to a specific goal necessarily influences the decision rules and the outcome of that decision. Choices therefore are influenced not only by goals—which may explicitly, or only implicitly, refer to risk—but also by the rules used to arrive at the final decision.

INFORMATION COSTS, UNCERTAINTY AND MIGRATION

There is tension between globalisation and technology changes, and the costs of enhanced decision making, especially given the nature of risk and uncertainty. A globalising world and modern means of transport and communication provide hitherto unprecedented opportunities for developing individual life and job strategies. However, an increasing array of opportunities makes decision making increasingly difficult. An individual may examine hundreds of alternatives and each may have dozens of attributes, such as type of job, place of work, volume of pay, availability and costs of travel, local amenities, etc. Given the massive information overload, it is impossible to assess all the alternatives and their potential outcomes. Moreover, high-consequence decisions, such as which job to take and which country to work in, are subject to substantial uncertainty and complexity. They often involve difficult trade-offs, while high financial and psychological costs could be incurred in order to reverse their outcomes. 'The decisions that matter most in life are often those that we are least prepared to make' (Kunreuther *et al.* 2002: 260).

Individuals have limited processing capacity (Simon 1955). There are some two hundred countries and virtually thousands of potential destinations for migration. For some locations and job types, abundant information is available, while for others this is scarce or inaccessible via formal information channels. The potential migrant does not have perfect information: instead, they have to cope with both information overload and incomplete information. No individual is capable of collecting and processing all the available information resources. He/she has to use heuristic strategies to limit the number of potential alternatives and costs related to their evaluation. The choice of heuristic strategy may reflect not only information cost optimisation, but also personal preferences and attitudes towards particular aspects of migration decisions. Some migrants may concentrate more on monetary costs/returns than non-monetary ones, while others may consider non-monetary factors to be more important.

While human capital theories implicitly consider risk, they have a major drawback in the assumption of perfect and freely available information on migration costs and return, and their associated risks. It is easy to see that once this assumption is dropped, potential migrants have to consider not only the costs and returns of particular migration alternatives (e.g. to stay or to go? where to go?) but also the costs of acquiring and analysing relevant information. But gathering information is demanding on time and sometimes on travel costs. The more severe the uncertainty about the potential outcomes of migration, the greater the motivation for collecting and analysing information.

Standard economic theory usually refers only to the costs of information acquisition. Stigler (1961) supposed that all relevant information can be obtained at some costs. At some point, the marginal utility of extra information is lower than the costs of information acquisition, and data collection is terminated. The founders of classical (micro)economic theory (Alfred Marshall, Adam Smith) had realised that the assumptions about perfect information were unrealistic, and that it was impossible to describe economic processes without all the necessary details. However, they lacked a methodology for describing decisions under incomplete information. The neoclassical synthesis in the 1950s and 1960s had a methodology (e.g. from Stigler 1961), but largely ignored considerations of incomplete information. Stiglitz (2000: 1443), for example, notes that:

> Mainstream economic theory, embodied in the competitive general equilibrium theory formalized by Arrow (1964) and Debreu (1959) simply ignored these considerations. There was the hope conveyed by Marshall's dictum, 'Natura non facit saltum,' that so long as information was not too imperfect, economies with 'almost perfect' information would look very much like economies with perfect information, close enough that the idealized models would suffice.

Complex choices may involve searching for very large amounts of information. The search for new information may be limited by: (a) the costs of information acquisition (at least in terms of time spent), (b) the cognitive capacity of a decision-maker (bounded rationality, [Simon 1955]) and (c) the previous knowledge, views and attitudes of decision-makers. Psychology highlights the importance of the personal traits and attitudes of a decision-maker in terms of information search. As indicated by our own research (Baláž *et al.* 2014), individuals' information needs may differ vastly. Individuals may look only for information they consider worthy, which depends on their attitudes and preferences. There is also a relation between the 'need for knowing' and information search. Individuals with a higher need for knowing devote more time to information search than those with low needs (Verplanken *et al.* 1992).

There are many different determinants of the need for knowing. The most important determinant may be the goal of reducing stress via solving the problems which generated the stress—that is, getting more information to reduce uncertainty. Uncertainty is usually a key stress factor. Van Zuuren and Wolfs (1991) assume that monitoring the environment and searching for information is highly correlated with problem-solving behaviour. Uncertainty, for example, can be a significant factor of stress and frustration. Acquisition of information may decrease uncertainty and relieve frustration and stress. Individual decision-makers use this to solve different sets of problems and have different motives for searching for particular types of information:

- need to find new information
- need for understanding existing information
- need to confirm relevance of existing information
- need for understanding and/or reconfirming one's beliefs and attitudes (e.g. looking for crime statistics to confirm a view that this is a dangerous country)

Information Overload and Imperfect Information in Migration Decisions

Migration researchers have acknowledged the problem of assuming complete and costless information. Fischer *et al.* (1997: 63) comment that:

> The absence of complete and costless information implies that any decision to 'stay' or to 'go' involves not only assessing and weighing different (known) migration alternatives. Potential migrants also have to worry about obtaining and selecting relevant information.

Unable to acquire all the information required or available, migrants are likely to make suboptimal decisions, that is, they make bounded rational decisions. Nevertheless, these are still rational decisions.

Rationalist models have limited explanatory powers when applied to real-life situations. In consumer research, field observations and experiments indicated that consumers spent much less time in the various stages of the decision-making process than was assumed by rationalist theories of consumer behaviour. Consumers also develop and adapt a diverse repertoire of strategies for deciding on and purchasing consumer goods. Some strategies involve relatively limited or no search for information.

If a set of alternatives and their attributes is too large, a decision-maker usually considers just a subset of the total information available. In our own research (discussed later in this chapter), for example, we presented 157 individuals with a set of 10 migration destinations and eight attributes attached to each destination (Baláž *et al.* 2014). The complete decision set therefore contained 80 information units. The participants were free to ask for as much information as they wanted, but the median value of information sought was only 31.6 information units. Only 8% of participants asked for all 80 information units. Therefore, most subjects consciously formed their decisions under imperfect information conditions.

There are different opinions about how individuals deal with imperfect information:

- By substituting average values for the missing information (Ganzach and Krantz 1990; White and Koehler 2004).
- By assuming that missing information is negative (Jaccard and Wood 1988; Johnson and Levin 1985). For example, if you have information that wages are high, but have no information on housing costs, the latter are assumed to be high.
- Missing information is assumed to be positive (Levin *et al.* 1985; Kivetz and Simonson 2000). For example, assuming there are no major health risks if no vaccination is required.
- Missing information is ignored (Sanbonmatsu *et al.* 1992; Simmons and Lynch 1991).

Garcia-Retamero and Rieskamp (2008) used an experimental approach to test these propositions. They found that averages are substituted for missing information where the latter is randomly distributed (that is, there is no reason to seek more targeted solutions). Negative or positive information is substituted where the missing information is conditionally distributed. For example, where data on the energy efficiency of a house are missing, it is assumed (negatively) there must be a hidden problem. In contrast, if data on the toilet are missing, it is assumed (positively)—at least in a developed country—that it is too obvious to mention. In short, negative information is assumed in the case of energy costs, and positive in the case of the existence of a toilet. They also found evidence that individuals simply ignored missing information when they considered it to be irrelevant. The authors' own research found some support for the abovementioned assumptions, and this is discussed later in this chapter.

The Decision-Making Process

The strategy adopted when making a decision depends on task complexity. The larger the numbers of alternatives and their attributes, the greater the demands made on the individual to process information. Attention is a scarce resource and there are also limits to individuals' memories and their computational capabilities. Tasks with high numbers of alternatives and multiple attributes may not lend themselves to a 'rational choice' strategy for most individuals. Instead, many complex decisions have to take place under 'bounded rationality' conditions (Simon 1955).

At some point, the marginal costs of information processing may surpass the benefits from a decision outcome. A satisficing solution may, therefore, be preferred to an optimal one. However, cutting the costs of information and (subjective) factor weighting via heuristic methods (short cuts) of information gathering and processing does, of course, pose its own risks. It may result in the choice of suboptimal (second-best, n-best) alternatives, unlike the optimal ones suggested by neoclassical approaches. Incomplete information may result in deciding on a suboptimal alternative (Maier 1985) and, for example, may result in return migration earlier than anticipated because of the poorly informed initial migration decision (Saarela and Rooth 2012; see Box 4.1). How much information is collected, and how different attributes of different destinations are weighted, depends on the potential migrant's risk aversion. Risk-averse individuals are likely to gather more information and to pay a 'higher risk premium' (in terms of information costs) than risk-neutral and risk-tolerant ones.

Box 4.1   Incomplete Information, Uncertainty and Return Migration

Saarela and Rooth (2012) analysed data on 7,729 Finnish male emigrants aged 25–55 years in Sweden for the period 1988–2004. A quarter of these emigrants had returned to Sweden by 2005. Migrants with zero or very low earnings were much more likely to return within a year than those with high earnings. Saarela and Rooth assumed that Finnish people returning within a year were likely to have had bad luck and/or incomplete information on job opportunities in Sweden. After controlling for age, education, marital and employment status, number of children, country of residence and first language, they found that uncertainty in the initial migration decision could be an important motive for return migration. The uncertainty arose from the incomplete information available when making the decision to migrate, as well as the inherent uncertainty of the future.

Source: Saarela and Rooth (2012)

The cost/benefit approach to complex choice making suggests that a decision-maker would favour strategies which offer a good ratio between the costs of the decision-making efforts and the accuracy of the result. The repertoire of decision-making strategies varies depending on individuals' particular experience and training (Bettman *et al.* 1998: 194). Most obviously, those with relevant expertise may be better able to recognise patterns of information distribution, thereby reducing the costs while also being better able to identify accurately the alternatives with better potential outcomes. Perceptions of complexity and framing may also impact on decisions. A decision-maker evaluates the format of the information, how it is presented and structured, its relevance, and the extent to which it is in tune with his/her beliefs, attitudes and existing knowledge.

Many decisions are complex and involve multiple alternatives, and most of the alternatives have multiple attributes. A prospective migrant, for example, might consider 10–15 alternative destinations, and these have a number of important features for the decision-maker, such as income, living costs, availability of family and friends, language and culture differences, climate and crime rates. Some attributes are compensatory but some are not. High income, for example, can compensate for living costs, but a nice climate is less likely to offset low earnings. Decisions involving high uncertainty and difficult trade-offs are especially complex. There are limits to the capacities of decision-makers to process information—they have limited working memory and computation abilities. Processing information about 15 alternative destinations with 10 compensatory or non-compensatory attributes is demanding. Hence complex decisions are taken under bounded rationality, and are informed by both existing, well-defined preferences (such as beliefs and attitudes) and preferences constructed during the process.

Research on complex decision making indicates that individual behaviour is highly adaptive to the difficulty of the task (Payne 1976: 367). A decision-maker first has to examine the demands on his or her information processing capacity. How many alternatives are there to evaluate and what are their potential outcomes? How many attributes exist for each alternative? Are all the attributes of the same importance, or are particular attributes dominant? Are particular attributes compensators: can a strong positive value for one attribute compensate for poor values on others? In other words, there are a large number of issues to be addressed, and many of these emerge during the decision-making process.

At the most basic level, complex tasks usually involve two-step decision strategies. At the first step, a decision-maker tries to reduce the complexity of the tasks and narrow the number of alternatives to an acceptable or manageable level. This is typical when using non-compensatory heuristics, such as elimination by aspect (Tversky 1972) or the lexicographic strategy (Fishburn 1974). The elimination by aspects excludes options which do not meet the cut-off value (the 'aspiration level') for the most important attribute. For example, when individuals state that they have removed from consideration all destinations where the nominal wage is lower than £2000. The lexicographic heuristic selects the option with the highest value on what is considered the most important attribute. For example, this could be that they will migrate, to the destination with the highest wage level, irrespective of its other attributes. The second step usually involves a more detailed inspection of the remaining alternatives based on their compensatory weighting. The compensatory strategies enable weighting; multiplying the values of the outcomes by their probabilities yields their decisions weights. The latter strategy may involve processes described by prospect theory such as loss-aversion and nonlinear transformation of probabilities (Chapter Two).

The simple notion of two-step decision strategies could be further disaggregated in terms of phases of information processing. Each phase may involve the use of heuristic rules which decrease the computational efforts required from the individual (Shah and Oppenheimer 2008). The first phase usually involves the identification of cues, that is, of relevant items of information. In the face of a large number of items, a decision-maker may opt for fewer cues and start with the more important ones. This is typical for the lexicographic heuristic and/or elimination by aspect heuristic mentioned above (see also Box 4.3). Considering fewer cues facilitates the comparison of alternatives. The second phase refers to simplification of the cue values and reducing the difficulties associated with storing them in, and retrieving them from, memory. A migrant need not memorise the exact values of living costs and wage levels. Instead, they may code countries as 'cheap or expensive', or 'high or low wage'. Reducing the complexity of information significantly reduces the computational demands of decision making.

The third phase involves assigning decision weights. Most complex decisions require some form of compensatory weighting. High wages, for example, may compensate for higher health risks and/or harsh climate. In many cases it is impossible to know the exact values of the weights allocated to particular attributes. Instead of numerical weights, decision-makers are likely to use emotional evaluations of cues (e.g. 'I would never work in a hot and humid country') and/or simplify weighting principles. Alternatively, the equal weighting heuristic, or tallying, assigns the same weights to all decision attributes, as a means of simplifying complexity.

In the fourth phase a decision-maker decides whether to integrate information from all cues or just from some of these. Satisficing can be an effective effort-reducing strategy. A migrant may set maximal cut-off levels for crime levels and/or health risks, and a minimal one for wage levels. They need not consider climate or life satisfaction levels in potential migration destinations as long as these satisfactory levels are achieved (see Box 4.2 on binary choice strategies). Destinations which do not pass these 'aspiration levels' are excluded from further inspection. A country which surpasses the cut-off levels on the most important cues may be considered 'good enough' if not optimal for migration.

Finally, some alternatives can be discarded before entering a decision set. Narrowing the set of alternatives helps reduce effort and focusses attention on the most promising choices. A migrant may use the recognition heuristic to choose a country with the best value on the attribute name recognition. Recognition is closely related to the image of an object (a destination, in terms of migration). Countries with positive and well-established images are usually preferred to less-known countries, even if they have similar levels of development.

The strategy used for acquiring information need not necessarily be identical to the strategy used when making a final decision. The information acquired is important for understanding the structure of the problem. The decision-maker confronts the newly acquired information with the information in their memory, as well as with their beliefs and attitudes. Moreover, the patterns of information acquisition can be impacted on by previous knowledge, images and beliefs. Therefore, the final decision reflects both external and internal information processing.

COMPLEX DECISION MAKING: A SLOVAK CASE STUDY

Following on from the general discussion of complex decision making, this section reports the findings of the authors' own research on complex decision making in relation to potential Slovak migration. Slovakia has one of the highest emigration rates in the EU, 27: 8% of the total labour force having worked abroad in 2012 (SOSR 2012). The research starts by initially considering simple binary choice, and then adding increasing complexity, comparing imperfect and complete information contexts, as well as the importance of images as sources of surrogate information (Baláž *et al.* 2014).

The research is based on a relatively large (for this type of experimental research) purposive sample of 157 advanced undergraduate students and postgraduates, and relatively recent graduates currently in employment in Slovakia. Their average age was 25. The interviews and experiments with each participant were time demanding, requiring about 2–3.5 hours. The sample contained broadly equal numbers of migrants and non-migrants, and each subsample had a broadly balanced gender composition. The experimental research methods were based on Active Information Search (information is provided through a dialogue with the interviewer, so the respondent never sees how the researcher has structured the data). Mouselab style information tracking was also used to record the information search processes used by individuals. The participants were asked to imagine they had been offered employment by a multinational company in any one of their ten international offices for a period of three years. Information was potentially available on eight items of information about these countries.

- Monthly net wage (after tax), in euros. These reflect actual wages in these countries but with additional wages being paid as a form of risk premium in countries where living and working conditions were challenging.
- Living costs.
- Climate.
- The crime rate.
- Health conditions.
- Language difficulty.
- Life satisfaction.
- Personal freedom and security.

Following Monti *et al.* (2009), the research assessed how individuals dealt with incomplete and complete information, and made decisions under different conditions of information availability. Stage 1 is a simple binary choice experiment, when individuals are presented with limited access to information for two unnamed countries. Stage 2 is a complex experiment involving unlimited information for 8 attributes for 10 countries—that is, when the participants are faced with potential information overload. They know the names of the countries, so these may act as surrogated information. In stage 3, they have complete information, being presented with full information for 10 unnamed countries—again a potential overload—but the country names are withheld in order that image does not play a role in surrogating information. Consecutive trails were held at each stage, using different sets of countries, in order to examine the consistency of information search and judgement formation.

Binary Choice for Unnamed Countries in the Face of Limited Information

Simulating the costs of acquiring data, participants could only request up to a maximum of one half of the total information available (four out of eight items). Further details are provided in Box 4.2. They were most likely to seek information on wages (86.1%) and cost of living (80.1%): in other words, more than four-fifths of the sample asked for information about the basic economic data that are central to (neo)classical economic models. However, noneconomic attributes, such as freedom, language and life satisfaction, were also important (see Figure 4.1). Non-migrants were more likely than migrants to seek information about language difficulty, wages and living costs. The focus of non-migrants on economic factors reflects their uncertainty about the returns to migration, while they asked about language because they lacked the experience of migrants in having lived abroad, and the challenges this represents in terms of learning a language. In contrast, migrants were more likely to seek information on climate and life satisfaction, perhaps because their experiences indicated the importance of these, while they were less concerned about the economic risks of migration.

Box 4.2   Identifying Strategies in Binary Choices

In the binary choice experiment, participants are presented with information on eight attributes for two unnamed countries. Limited access to information is simulated by only being allowed to request up to 50% of the available items (4 ut of 8), necessitating they request the information they consider most relevant. Countries are not named in order to avoid image influencing behaviour. In this particular experiment, data for Country A represented Serbia, while Country B was Paraguay. Information limits did not allow for consistent evaluation of all attributes. The participants could only use heuristics based on selective evaluation of attributes, such as satisficing, a disjunctive strategy, and/or recognition heuristic.

Countries have significantly different mixes of migration risks, costs and profits. For a potential Slovak migrant, Country A would represent an undemanding and medium-return (real income) destination. Country B offered a better ratio of wages and costs (monthly living expenditure), but also higher health risks and very high crime rates. To provide one example, participant no. 46 asked only for (1) wage levels, (2) life satisfaction levels and (3) monthly living costs, that is, for only 3 information units out of the total 8 that were possible. Country B passed the satisfaction cut-off levels on the abovementioned migration attributes better than Country A. The short retrospective verbal protocols revealed that the participant identified both wages and living costs as major factors, rather than considering one factor to be of overwhelming importance. The decision making strategy was satisficing.

Table 4.1   Example of binary choice: participants can request up to 50% of the available information

Variant 4	Health risk	Language	Freedom and security	Wage (EUR)	Crime rate	Life satisfaction	Climate	Minimal monthly living costs (EUR)
Country A	Average	Easy	High	1100	Low	Medium	Mild	350
Country B	High	Medium demanding	Medium-high	1600	Very high	Medium-high	Warm	250

Source: Authors' research

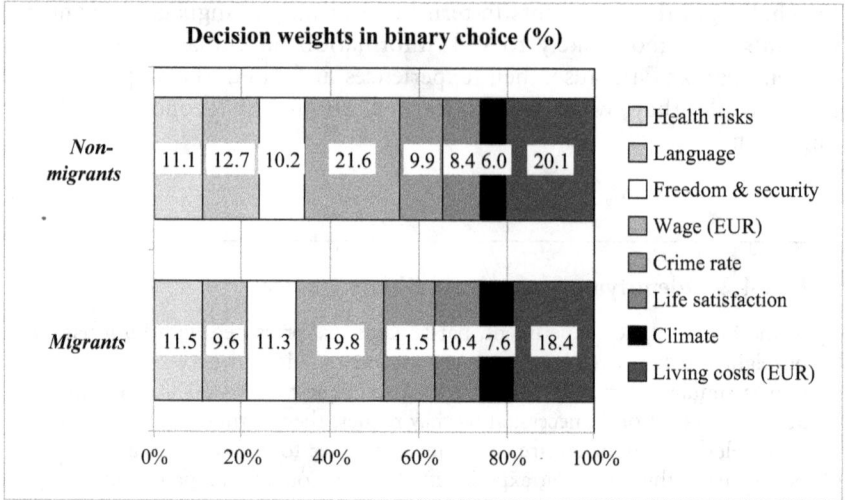

Figure 4.1   Decision weights in binary choices (%)

Source: Authors

Complex Choice, Unlimited Information and Image

In the second stage, the participants in the experiment could acquire up to 8 items of information for 10 named countries. That means they could obtain a maximum of 80 items of information under these conditions of unlimited access to information. This approximates to a complex choice for an individual who is faced with large volumes of unstructured information. For potential migrants this represents potential information over-load which—in a different context—is known to influence decision making (Lee and Lee 2004). On average, they sought information for 45% of the 80 units potentially available (see Box 4.3). Very few participants sought all the available information, and most sought to reduce the information load significantly.

Box 4.3   Identifying Strategies in Complex Choice Experiments

In a complex choice environment participants were allowed to search for as much information as they wanted on 8 attributes for 10 countries, that is, they could ask for anything from 0 to 80 units of information. Countries were named, signalling surrogate information in terms of their images. At the end of the task, participants ranked the 10 potential destinations on a 10-point scale, ranging from 1 = the most likely country for migration, to 10 = the least likely country for migration.

As an example, we consider participant number 92. In step one, the participant asked for information on (1) wages for all countries. He used the Elimination by Aspect Strategy (EBA) to exclude countries with low nominal wages: Pakistan, Croatia, Turkey, Chile, Guatemala and the Czech Republic. The EBA heuristic narrowed the choice of alternatives to Japan, Denmark and Ireland. In step two, the remaining alternatives were further inspected in terms of selected important attributes, namely (2) living costs and (3) climate. This type of information processing is typical for the 'majority strategy', which selects the option with the highest number of dominant attribute values. Japan was ranked first, Denmark second and Ireland third as the most likely country for labour migration. The lowest ranks were assigned to some less developed countries. The short retrospective verbal protocols revealed that the participant combined information acquired during the experiment with images of the countries in order to rank potential migration destinations.

Table 4.2 Example of information search strategies in complex choice: order in which information is searched for

Country	Health risks	Language	Freedom & security	Wage (EUR)	Crime rate	Life satisfaction	Climate	Living costs (EUR)
Japan				1			3	2
Denmark				1			3	2
Pakistan				1				
Croatia				1				
Guatemala				1				
Chile				1				
Czech Republic				1				
Ireland				1			3	2
Turkey				1				
Kuwait				1				

Source: Authors' research

The total information sought decreased modestly over the four trials indicating that there was a learning effect based on their experiences. However, the learning effect was relatively limited because individuals had already expressed what they most valued in the first trial, and this did not substantially change. Preferences were stronger than the learning effect, as Bettman *et al.* (1998) have argued. They are more present and persistent than constructed (Warren *et al.* 2011), but both effects were observed. Migrants consistently (across the four trials) asked for less information

Decision weights in complex choice (%)

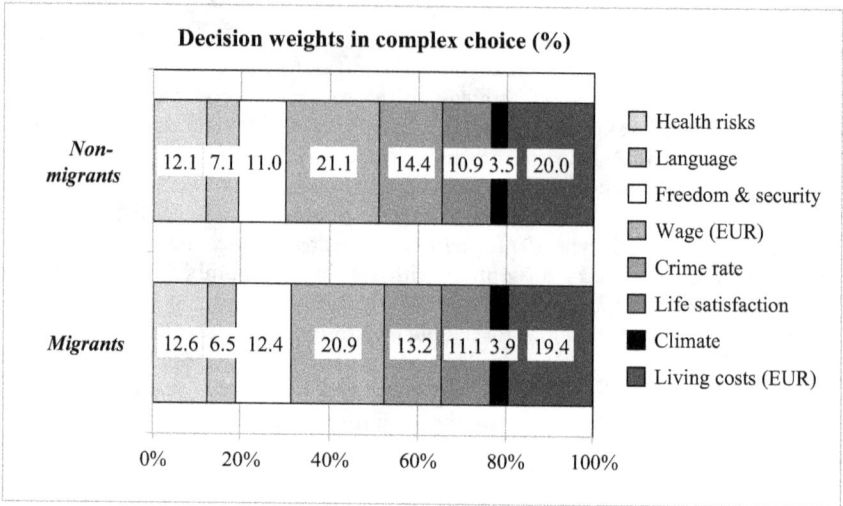

Figure 4.2 Decision weights in complex choices (%)
Source: Authors

than non-migrants, although the difference was insignificant. However, the learning effect was greater for non-migrants than for migrants: this is consistent with migrants' greater tacit knowledge and experience, leading them to seek more information initially.

The main factors that participants sought information about were again economic—wages and living costs—followed by crime, health risks, freedom and democracy, life satisfaction, language and climate (see Figure 4.2). These rankings were broadly similar to the simple binary experiment, but language was less important, as the name of the country surrogated information about the difficulty of that country's language. In general, compared to the binary experiment, greater weight was attached to those elements with a higher degree of uncertainty because of the lack of information surrogated by their names: this is particularly important for noneconomic factors. Non-migrants were again more likely than migrants to seek information about wages and costs, but not about language, for the reason stated above.

The images of the countries clearly had a significant influence on decision making—that is, on how countries were ranked. Image had high, significant correlations with both knowledge of the countries and their rankings—it was higher than the correlation between knowledge and ranking. Effectively, the names of the country signal information to potential migrants. There was again a learning effect, and over the four trials there were increases in the correlations between both image and knowledge with the rankings of the countries. Migrants and non-migrants had broadly similar correlation coefficients.

In summary, the experimental research indicates that individuals learned how to manage potential information overload by selectively seeking information. Migrants, drawing on their knowledge and risks of having lived/worked abroad, tended to ask for less information than non-migrants. Image seemed to be more important than knowledge in decision making, but image—including the image of risk—was based on knowledge, some of which was accumulated during previous migrations. Although there was some learning across experiments, this was far less important than initial knowledge and image.

Complex Choice, Complete Information and Image Suppressed

The final experiment assessed decisions made when images are withheld, that is, the country is unnamed, while also providing complete information to the participants in a clearly structured format. Some of the trials held in the previous stage are repeated for the same countries, without the participants knowing this, but their names are withheld, and the data are reorganised to minimise memory recall. The main focus was on estimating the decision weights, or the relative importance of the different factors, in the two complex experiments. A range of methods were used to calculate these, reflecting differences in the two data sets.

There is a striking similarity in the weights in the two stages. The dominant attributes in potential migrants' decision making about potential destinations were economic: wages and living costs (see Figure 4.3). Health risks

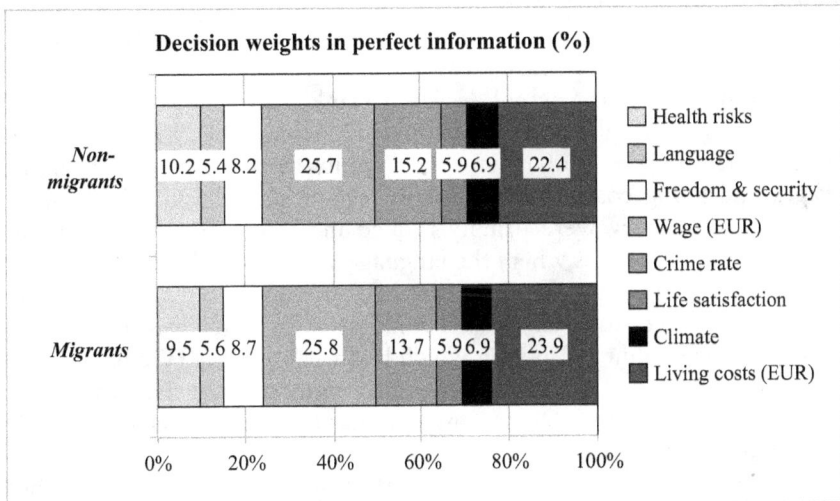

Figure 4.3   Decision weights in perfect information conditions (%)
Source: Authors

and crime rates were also important. There were no significant differences between migrants and non-migrants.

The comparison of the decision weights produced by the same participants for identical sets of countries under the unlimited (but incomplete, in practice) and complete information conditions were revealing. There were significant differences in *all* the attribute weights between the two stages of the experiment. The relative importance of wages and living costs was greater under complete information conditions, while the other non-economic factors were mostly relatively weaker. Nevertheless, the individual participants' preferences were broadly consistent when seeking information, which indicates a strong underlying component in judgement formation about potential destinations—and about the implicit risks associated with these. Arguably, the relatively high importance of living costs in the judgement formation stage is a manifestation of loss aversion (Tversky and Kahneman 1992).

The comparison of identical sets of countries in the two experiments also facilitated assessment of the influence of images. The average country rankings were broadly similar irrespective of whether the names of countries were known. However, there were some differences. The Czech Republic and the UK were less attractive when participants knew the name of the country—perhaps because these are well-known destinations for Slovaks, and they understand the positive and negative risks of migrating to them, that is, they have more realistic images of these particular destinations. Finland, Taiwan and Oman were *more* attractive when the country name was known, and this may reflect an image advantage as few Slovaks had any real (tacit) knowledge of these countries. Iran was more unattractive when its name was known, reflecting its weak image in Slovakia. These findings provide evidence of the importance of image in terms of migration risks.

When it came to the actual decisions (rankings of destinations), most participants generally first ranked the developed EU Member States (Denmark, Norway, Finland, Sweden) which had the highest nominal wages and living costs. When the UK, Ireland and the Czech Republic were included, they were usually ranked in second or third positions: these were all countries where the language was less likely to be a barrier, and where existing large migrant networks effectively reduced the potential costs of migration. The lowest rankings were associated either with less developed countries in Asia which had weak images (Vietnam, Laos, Cambodia, Pakistan) or with oil-exporting countries that the participants had little knowledge of (Qatar, Oman)—that is, where image outweighed potential economic gains.

Migrants assigned higher rankings than non-migrants to established migration destinations, such as the UK, Czech Republic and Ireland, and had more negative opinions of migration to distant and/or less developed destinations such as Laos, Guatemala, Chile and Argentina. This may be because of their more realistic view of migration based on their previous experiences. Interestingly, the differences between migrants and non-migrants were greater when the county names were known, that is, when image had a stronger influence. Non-migrants' preference for more exotic destinations may reflect their lack of experience of the real risks, positive and negative, of migration.

CONCLUSIONS

The limitations of the assumptions of perfect information are well known (Stigler 1961; Akerlof 1970; Stiglitz 2000), and this has implications for the role of risk in migration: where knowledge ends, risk begins. However, although this has obvious application to migration studies, there has been relatively little research that has sought to understand how migrants deal with different information conditions in terms of complex decision making. The authors' work on complex decisions reported above is an attempt to illustrate the contribution of complex decision-making models to understanding migration.

The findings of the authors' research provide some support for the neoclassical focus on the economic determinants of migration, notably wages and living costs. However, it also demonstrates that noneconomic considerations are important, such as living standards, health and crime. In making decisions—which includes at least implicitly considering risk—the individual participants draw more on their incumbent preferences than the preferences constructed during the decision-making process. However, although there are relatively strong and stable underlying preferences (Kivetz *et al.* 2008), there is also some evidence of learning and of preference construction. The incumbent preferences are strongly influenced by images which signal information (Stiglitz 2000) and existing knowledge of migration destinations, including risks.

Migrants face information overload and imperfect information conditions when seeking to make decisions. Experimental research methods indicate that very few participants used all the available information, with less than one half being requested on average. This approximates the real world of imperfect information (Stiglitz 2000) that migrants encounter. There was also a modest learning effect, as the amount of information sought decreased across four successive trials. Image and existing knowledge help reduce information overload, through surrogating some information. Migrants already possess some tacit knowledge about migration and deal with information overload more efficiently than non-migrants.

To date, and to the best of our knowledge, there has only been one substantial attempt to apply complex decision-making concepts to migration. Moreover, we acknowledge that there are important limitations to observing situations in laboratory experiments (Birnbaum and LaCroix 2008). However, we also contend that experimental research methods do provide a means to obtain valuable insights into how individual migrants contend with different information conditions—and this lies at the heart of risk considerations. There is therefore a need for further research to evaluate the contribution of this approach, including its application to different groups of migrants.

5 Households, Risk Diversification and Migration

HOUSEHOLDS AND RISK DIVERSIFICATION

Economists tend to treat migration as individual decision making, whereas sociologists are more likely to consider the role of families (Mincer 1978). There are several reasons for this economics perspective, including the argument that the utility of the household head is the main determinant of the household's utility (Becker 1991). There are, of course, a number of gendered and other assumptions implicit in this perspective, which clearly fails to capture the complexity of household decision making (Himmelweit 1998). In contrast, sociologists contend that households are fundamentally different to individuals as decision-makers, and there is a need to focus on internal social relationships within the household or family. To do otherwise is to treat the household decision-making unit as a black box (Burgoyne 1995).

The consumer decision making literature has increasingly recognised the roles of households or families (Lackman and Lanasa 1993). One take on this was the emphasis on the different roles adopted by family members. For example, Engel *et al.* (1990) identified five specific roles: gatekeepers, influencers, decision-makers, buyers and consumers. The gatekeepers initiate the family decision-making process, being the first to recognise a particular need; the influencers are the family members whose opinions are sought; the decision-makers are those who eventually make the decision; the buyers physically purchase the product; and the consumers are those who benefit in terms of utility derived from the purchase. It is possible to think of household migration decision making in these terms—with one or more of these different roles being assumed by particular individuals, depending on the distribution of power and practice within the household. This offers interesting perspectives in terms of considering the extent to which the migrant is the benefactor, the decision-maker, the initiator, the opinion provider, or simply the 'purchaser' in the sense of being the individual who implements the family decision. However, it does not provide an understanding of risk and decision making, and the analogy with consumption is not ultimately compatible with how migration interacts with labour market participation.

For the latter, we can turn instead to what has become known as the new household economics, notably the work of Mincer (1978).

Mincer (1978) argued there was a need to look at net family utility rather than individual utility resulting from migration. His starting point was the work of Long (1975), which demonstrated that migration was a family decision (excepting households composed of single persons). Even if the lead migrant could anticipate an increase in income from migration, families could be deterred from moving if the total economic and noneconomic costs to the family exceeded these gains. This was especially likely when children were of school age. Building on this, Mincer (1978) developed a more elaborate model based on the notion of ties. Ties exist when individual (private) gains from migration have different signs for the different household members—that is, they are positive versus negative for different individuals. The net loss experienced by the tied mover must be smaller than the net gain of the spouse for family migration to occur. The nature and extent of these ties are related to income, so that two-earner families are more likely to be deterred than single-earner families. And divorced and separated individuals have the highest migration rates—higher even than single persons, whose mobility, on average, may be mediated because of their membership in their parents' households. Empirical testing of the model indicates the highly gendered nature of migration. Migration tends to reduce men's unemployment but increases women's because they tend to be tied, having to give up their jobs when migrating with their partners, and are faced with uncertain employment prospects in the destination. However, there is evidence that lead migrants make some attempt to ameliorate this distribution of the costs and risks of migration; for example, unemployed men with working wives tend to search longer locally for new job opportunities than do men whose wives are not working. Mincer (1978) proceeds to argue that there are also implications for divorce: 'By imposing negative private externalities on at least one of the spouses, family location decisions can be a challenge to family integrity': when the externality exceeds the gain from being married, then the marriage may be dissolved. Risk is only implicit in Mincer's (1978) work. As with human capital theories, the risks attached to migration are implicitly priced into the returns required to trigger migration, but they are not explicitly analysed as was noted earlier in relation to some of the neoclassical approaches, and especially in the behaviouralist approaches. However, one specific theoretical approach does address the issue of risk at the household level, the 'new economics of labour migration'.

NEW ECONOMICS OF LABOUR MIGRATION

The NELM theory, or the new economics of labour migration (Stark 1991), assumes that migration decisions are made not at the individual level, but at the household level. It addresses the notions of risk, market imperfections, and the role of migration in the diversification of income and risk. It is a significant advance on neoclassical theory, which only recognises migrant remittances as having unitary income effects on the income of the recipient households as a whole (Massey *et al.* 1994): 'Because risk is disregarded and all markets are assumed to be complete and well-functioning, production decisions are presumed to be independent of household budget constraints and other sources of income'. In contrast, the NELM approach considers that households minimise risks in multiple markets, some of which function imperfectly (Stark and Bloom 1985).

The origins of the NELM approach lie in research on rural-urban migration in less developed economies, where it was recognised that local markets were imperfect, so that access to particular labour markets 'is often barred by constraints in capital, commodity, or financial markets' (Stark 1991: 4), generating high levels of risk for households. Stark, and his associates, considered that migration was instrumental for households in combating risks arising from imperfect or underdeveloped markets. The relative absence of these imperfect market conditions in more developed economies means that the NELM model is usually considered to be specific to less developed economies. Of course, state policies can seek to mediate risks via, for example, unemployment insurance or crop price guarantees, thereby influencing migration, but this is limited because migrants face uncertainty rather than risk, and it is not always possible to insure against uncertainty (see Chapter Seven). States may also lack the resources to provide such insurance. Given the lack of commercial or state provided insurance, the NELM models consider that households invest in the costs of migration for one or more of their members in expectation that they will provide remittances which can be used to mediate local market failures, such as bad harvests or other natural disaster.

Vullnetari (2012: 202) provides an illustration of the role of migration in rural areas in Albania, which faced strong competition from Greek and Macedonian agriculture, as well as poor quality infrastructure. She quotes one of her interviewees, Berti:

> The apples were grown this year, but before harvesting they were all destroyed by some sort of disease. So people don't feel like working here. They say: 'It's better for me to migrate abroad, I know that I am working eight hours a day, but I know that at the end of the day I will come home and have money in my pocket.' Whereas here, you have to wait for a year to be able to get any money, if at all.

The NELM model sees migration as a means of providing insurance against such risks via diversification of income sources. Before proceeding further in our discussion of NELM, it is useful to consider the difference between insurance and other options, as indicated by Roberts and Morris (2003: 1253):

> An option is a contract which entitles, but does not oblige, its owner to a particular course of action. Options expand opportunities, and so are valuable. . . . Because options can be used to manage risk, they are often confused with insurance. . . . Insurance is thus a type of option—one that is concerned only with the downside of an asset's distribution of potential returns, or 'hazard'—while options more generally can take advantage of the upside, or 'opportunity'. In both, the potential for loss is fixed, but with insurance the potential for gain is limited to the value of the insured asset, while with an option the potential for gain can be unlimited.

NELM focusses on migration as a form of insurance, in the absence of a functioning insurance market in underdeveloped migration, although arguably, given that it also has potential for upside gains, this perhaps should be thought of as an option. Putting aside this issue and using the conventional terminology, the main dynamic is the provision of insurance against possible downturns, rather than considering migration to be a source of unlimited potential gains. This is specifically addressed by Massey *et al.* (1993: 436):

> Unlike individuals, households are in a position to control risks to their economic well being by diversifying the allocation of household resources such as family labour. . . . In developed countries, risks to household income are generally minimized through private insurance markets, or government programs, but in developing countries these institutional mechanisms for managing risk are imperfect, absent or inaccessible to poor families, giving them incentives to diversify risk through migration.

The key point is that the household risk reduction strategy is based on diversification, which not only considers the returns from different occupations—that, for our purposes, are in different economic spaces—but also the correlation of these returns (Stark and Levhari 1982). While an individual migration act might be highly risky, such as undocumented migration to precarious employment in another country, the risk to the household's income would be reduced if the returns in these different labour markets were uncorrelated (Roberts and Morris 2003). There is an important point of divergence here with human capital theories. The latter imply that the earnings differential between the places of origin and destination, which implicitly take into account differences in risks, is the driving force of migration. In contrast, NELM contends it can be logical for individuals to migrate to areas with lower incomes, if the outcome is a reduction in total household risk: 'Income is not a homogenous good, as assumed by neoclassical economics. The source of the income really matters' (Massey *et al.* 1993: 439).

An essentially rationalist approach, NELM assumes that households are the effective decision-making units, with decisions being taken about the distribution of the employment of the household members as a whole. The household—any household—faces risks and uncertainty in relation to future income streams. They simply cannot know what will happen to their current income sources in the future, and this is especially challenging where the income sources are heavily concentrated in a specific activity in a specific economic space, notably agriculture in a rural economy in a less developed economy. In such circumstances it is rational to seek to diversify income so as to reduce the risk and uncertainty that the household is exposed to. The migration of one or more family members to different labour markets or economic areas, whether internal or international, is one means of diversification. Risk is likely to be more diversified where that migration is international, on the assumption that international labour markets are less correlated than inter-regional markets, even though globalisation and increasing international economic integration are challenging this assumption. In effect, the migration of some family members, while others remain in the home economic space, is a means of diversifying the sources of incomes as a form of insurance against perceived income uncertainties.

Remittances have a double function in terms of risk diversification. They are, of course, a source of income to cover basic consumption needs, and effectively 'insure' consumption against threats to this from downturns in agricultural incomes. But they are also a means to secure capital in order to invest in the farm, and transform or diversify its production, so as to reduce exposure to risk. This is capital that cannot be secured locally in imperfectly functioning capital markets. In other words, 'migrants play the role of financial intermediaries, enabling rural households to overcome credit and risk constraints on their ability to achieve the transition from familial to commercial production' (Taylor and López-Feldman 2007: 3; see Box 5.1).

The insurance, or shared costs and risks, argument also works in the reverse direction. The household may provide a form of insurance for the migrant—a place to return to, and a shared household income in case the migration experience is economically unsuccessful. An important assumption in this conceptualization is that there is an implicit social contract between the migrant and the other household members that he/she will provide them with remittances (Lucas and Stark 1988), and will be able to benefit from the use of those remittances after returning. In economic terms, this assumes the existence of a non-separable model (Taylor and López-Feldman 2007). The assumptions underlying this theoretical perspective are of course questionable, including the persistence of loyalty to the shared household income, or non-separable model, especially when the individual migrants start their own families. There is also the role that imperfect information plays in risk pooling arrangements. Even

---

Box 5.1 Household Migration, Risk and Production Frontiers

Taylor and López-Feldman (2007) express the economic relationships between migration, insurance and production in a formal model, where the household has two possible production activities. It has a choice whether to invest its fixed resources, such as land, in either low-return or high-return activities. All else being equal, it would be expected to specialise in high-return activities. However, in practice there may be constraints on investing in high-return activities. Given the absence of effective insurance markets, a risk-averse household may be unwilling to invest more than a certain amount of its fixed resources in higher return, but relatively risky, activities. In these circumstances, selective family migration provides a means to relax the constraints on the household's credit or liquidity constraints. Remittances 'could provide income security by promising to remit in the event of a crop failure or other adverse shock, functioning de facto as a household insurance policy' (Taylor and López-Feldman 2007: 3). This would in effect shift the production possibility frontier by increasing the productivity of fixed assets such as land.

'Because the relative influences of migration on liquidity, risk and labor constraints are unknown, the overall impact of migration on total household income is ambiguous' (Taylor and López-Feldman 2007: 3). However, in the non-separable household model, the impacts are likely to be nonzero, in the face of capital and/or human capital constraints. Migration and remittances could increase output in the high-return activities by providing a means to relax these constraints, but this could also have a negative impact on the low-activity sector (for example, through loss of labour via migration). However, in the most positive outcome, migration and remittances may also loosen constraints on technology and fixed inputs, such as land and equipment, allowing increases in productivity in both sectors by shifting the production frontier.

Source: Taylor and López-Feldman (2007: 3–4)

---

if there is strong commitment to this contract, it may be difficult to enact consistently because of asymmetries in information amongst the family members; these are likely to increase with distance (Rosenzweig and Stark 1988).

There are also implied assumptions that the migrants may eventually return to participate in some of the utilities resulting from their transfers of income (Taylor 1992). Galor and Stark (1991) expand on this point, arguing that the social contract between migrants and the other household members may explain why individual migrants continue to work abroad to minimise risks to future household income, even when their current levels of savings would seem to offer the possibility of returning to a more prosperous future. For example, Dustmann (1997) has demonstrated that migrants jointly determine their savings levels and duration of their migration when faced with uncertainty. The interrelationships amongst these are dynamic, and are adjusted during the course of migration. Migrants are also more likely than non-migrants to accumulate precautionary savings.

The new economics of labour migration also points to the importance of social and family networks, one function of which is to help secure jobs in the destination. This can decrease uncertainty about employment and income in the destination, thereby enhancing differences in the levels of risk in the two markets—in effect, further reducing their correlation. The relationships between networks and risk are discussed further in Chapter Seven.

NEW ECONOMICS OF LABOUR MIGRATION: EMPIRICAL EVIDENCE

The NELM theory has attracted considerable research attention, and it has been tested empirically in a number of different contexts. Massey (1990) concluded that most of the evidence is supportive of the notions of risk diversification, that is, of the role of migration as a source of income diversity, risk reduction and provision of a form of insurance, while also allowing households to overcome barriers to accessing credit and capital. The notion of co-insurance is also evidenced by the work of Lucas and Stark (1985). Hamid (2007) provides a more recent review of the empirical evidence and, similarly to Massey, also concludes that quantitative and qualitative research demonstrate that remittances do reduce income uncertainty and provide sources of investment for families in rural areas in, amongst others, India, Peru, and Mali:

> NELM studies also emphasize how migrant and household remain linked through processes of 'coinsurance', whereby the household collectively, or members of the household, may fund initial movement and provide a kind of insurance against transitory income shocks whilst the migrant is getting settled, in return for the migrant providing equivalent insurance against unforeseen income shocks at home once they are established abroad.

Black *et al.* (2011a; 2011b) also consider that there is strong evidence in sub-Saharan African dry lands, subject to significant climatic variability, that migration is an attempt to secure household wellbeing via livelihood diversity. This is particularly marked for pastoralism and fishing, which are especially vulnerable to variability and instability.

There has been particular research interest in the effectiveness of co-insurance in the face of systematic risks, or exogenous shocks. In an overview paper, Kapur (2004) analysed the relationship between remittances and levels of private consumption in the years prior to and after economic crises. The latter were defined in terms of an annual reduction in income of 2% or more in a particular year. Kapur found clear evidence to support the co-insurance argument that remittances would increase following a macro-economic crisis. The evidence is, of course, limited, in that it does not indicate the uses made of the remittances, or the link between the increase in the remittances, and the extent to which the recipients had been negatively impacted by the economic downturn. However, it does provide a broad international overview of generalised relationships between economic conditions and remittances which is supportive of the risk diversification and co-insurance thesis. Box 5.2 updates some of the evidence relating to the role of migration in mediating macroeconomic-related risks.

Box 5.2   Remittances as a Factor of Decreasing Macroeconomic Risks

Globally, total remittance inflows surged from $1.9b in 1970 to an estimated $483b in 2011 (World Bank 2012). The growing importance of remittances in the global economy is also indicated by their increasing shares in world GDP. The share increased from a mere 0.07% to 0.72% in the period 1970–2011. In some countries, remittances generate a substantial share of total income. By 2011 remittances as a percentage of GDP were of the highest importance for some of the post-Soviet economies (Tajikistan 31.0%, Moldova 23.2%, Kyrgyzstan 20.8%), Asian countries (Nepal 20.0%, Lebanon 19.6%) and Latin America countries (El Salvador 15.7%, Jamaica 15.2%, Honduras 15.0%). As for the OECD Member Countries, remittances as a percentage of GDP were relatively important for Belgium (2.3%) and Slovakia (2.1%).

Remittance flows are countercyclical in relation to downturns and crises in the countries of origin (Mohapatra and Ratha 2012: 23). As for the crises in the destination countries, flows of immigrant workers tend to be reduced at such times, but there is little evidence of higher rates of return migration (Green and Winters 2012: 52). Immigrants tend to be more affected by economic crises than indigenous workers in destination countries, but most do continue working there and sending remittances to their home countries. In times of crisis, migrants may move between industries but continue to send remittances. Migrants are also accustomed to reducing their personal expenditures in order to maintain their remittances. This explains why remittance flows to developing countries are more stable than capital flows—and is a means of mediating the risks to households in the countries of origin of the migrants.

Countries with large emigrant communities can also benefit from geographical diversification amongst their workers abroad. This serves to diversify the income from remittances in terms of regional origin and exchange rates—that is, it increases the probability of sourcing remittances from labour

markets with uncorrelated risks. Geographical diversification of migrant communities essentially performs the same function as diversification of risky
assets in financial portfolios.

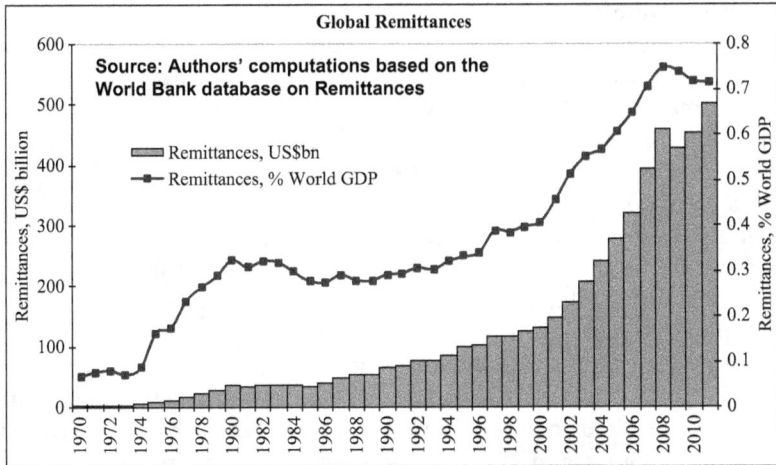

Global Remittances

Source: Authors' computations based on the
World Bank database on Remittances

Remittances, US$bn
Remittances, % World GDP

Sources: Author's summary of cited references.

A more detailed picture of the role of migrant remittances in mediating
household risks is provided by individual case studies. Lillard and Willis
(1997) found that the levels of remittances sent to Malaysian families are
sensitive to the current and prospective income available, while Rosenzweig
and Stark (1988) found that remittances to rural households in India also
vary inversely with agricultural profits. Thai remittances are higher when
the receiving household's income is lower (Miller and Paulson 1999). In
another study, Miller and Paulson (1999) have demonstrated that there is an
inverse relationship in Thailand between remittances and income levels,
probably reflecting changes in agricultural production in response to rainfall variations.

Suleri and Savage's (2006) study of Azad Kashmir in Pakistan is particularly instructive. This is an area where migrant remittances are the primary
source of income for most local households, and where there is clear evidence that remittances have significantly increased their capital assets,
including the ownership of land and housing. The region suffered a devastating earthquake which led to the collapse of communications, and severe
damage to housing and other infrastructures. Most households in the region
were affected, irrespective of whether or not they were in receipt of remittances. However, Suleri and Savage (2006) provide evidence that those
households which had been in receipt of remittances had been less vulnerable to the economic impact of the earthquake, presumably because they

possessed more reserves, in terms of both liquid and fixed capital. More-over, their household economies recovered far more quickly, once remit-tance flows had been re-established. They had access to sufficient resources to start repairing and rebuilding their houses relatively quickly, and to pay for private health care. In contrast, those households which could not call on the co-insurance of migrant remittances either had to rely on public health care or sell some of their scarce capital assets in order to pay for pri-vate treatment. However, there were spillover effects between those who were and were not co-insured, because the expenditure of remittances was instrumental in restoring local markets, and providing jobs for local work-ers in repairing houses.

A number of other studies have provided evidence that migration reduces the risks to household incomes in rural areas and the constraints on invest-ment, including Taylor (1992) and Benjamin and Brandt (1998). Taylor and López-Feldman's (2007) research in Mexico also provided evidence of the role of remittances in loosening risk constraints on household invest-ments. In broad terms, they found that the enormous increase in migration to the USA resulted in substantial remittances which both increased the per capita incomes and the land productivity of recipient households. Their work is particularly important in indicating that substantial pro-portions of the remittances were allocated to production as opposed to consumption.

It is, however, important to take a more cautionary view of the overall effects of remittances. One contentious issue is whether migration—whatever the impact of remittances—may result in labour shortages in the home region, leading to decreases in production. In other words, is there a high price to pay for co-insurance? Rozelle *et al.* (1999) and Taylor *et al.* (2003) provide evi-dence that in rural China, while there were positive remittance effects on production, there were also negative effects in terms of labour market losses. At some point, the potential losses in production in the traditional sector may become a deterrent to migration, and at the very least they may reduce the total net gain to the household. However, Taylor and López-Feldman (2007: 5) warn that:

> The incentive to invest in production activities, with or without migra-tion, is likely to depend critically on other variables, including access to markets for inputs and production in migrant-sending areas. . . . In short, the influences of migration and remittances on income and pro-ductivity in migrant-sending households are complex and cannot be signed a-priori.

While their overall findings suggest that rural Mexican household incomes and land productivity were increased by migration to the USA, it takes a number of years for these positive effects to be fully realised. Moreover, the extent of these positive impacts is of course dependent on total household assets, especially land ownership, and the extent to which households are able to tap into other local sources of income. They conclude that:

> Non-migrant households are positively selected into a nonmigration status. There is no evidence of positive sample selectivity bias for the migration-treatment group, however. This implies that if migrant households were suddenly deprived of migration, their expected incomes would be lower than those of otherwise similar households without migrants.

In other words, the simple two-production-sector model is necessarily a simplification, and that migration is only one of a number of risk-diversifying, or investment-funding, strategies available to households. Moreover, households with more limited resources are more likely to utilise migration as a source of risk diversification.

Another important study (Halliday 2006) has used panel data to investigate the relationship between exogenous economic shocks and migrant flows from El Salvador to the USA. It offered evidence that this migration is partly determined by prevailing economic conditions in El Salvador, and that migration provides households with a risk-management strategy. The average probability that a household sends members abroad is some 25% lower when there is an absence of economic crises, while remittances grow by 40–60% following major agricultural shocks. The most interesting finding was that the major 2001 earthquake did not result in an overall short-term increase in migration or an increase in the proportion of households with migrants. Halliday (2006) speculates that there may be two main reasons for this. First, that households needed the labour of all their members to help deal with the consequences of the earthquake. There was a short-term need for household labour, effectively providing a buffer to counter the effects of the earthquake. And, secondly, that the earthquake exhausted the household resources required to support migration—migration does require minimum material and other resources. The findings were, however, to some extent ambiguous. Higher income households were generally more likely to have migrant members, providing evidence of the importance of resources. However, all households were similarly affected by the earthquakes—so it was not a lack of resources which led to the overall reduction in migration levels.

CONCLUSIONS: THE ADVANTAGES AND LIMITATIONS OF THE NEW ECONOMICS OF LABOUR MIGRATION

The main advantage of the NELM approach is that by recognizing households as the effective decision-making unit, at least in rural areas in less developed economies, it acknowledges the potential implications of the distribution of household labour across different labour markets. This opens the way to understanding that income is not a simple unitary return to labour because there are different levels of risk attached to working or producing in particular markets. It is only a short step from this to understanding that households may seek to diversify their income sources in order to reduce exposure to risks in particular markets. Moreover, the remittances provided by the migrants may provide the additional resources that make households more willing and able to take risks by investing in new forms of production. Remittances, in effect, provide an alternative to the formal risk insurance systems (both state welfare and private market) that are incomplete or absent in these economies. It is particularly important in this book because it places risk explicitly at the heart of the migration decision-making process. There are, however, a number of limitations of NELM that need to be noted.

Firstly, as Engel and Strasser (1998) emphasise, both the exposure to risk and the response to this is partially structurally determined. Minimum resources are required to migrate, even if the advantages of migration as a means of diversifying against risk are recognised. The poorest do not tend to migrate. But it is also true that desperation may drive migration, even in the face of considerable risks. Vaughan and Dunton (2007) express this in terms of how poverty leads individuals to minimise the perceived risks and raise risk thresholds. In migration terms, it can be understood as an argument that extreme poverty may cause individuals to be willing to accept high levels of risk. This is not because they are more risk tolerant, or feel more competent to manage these, but because the risks of staying may be considered to exceed the risks of migrating, so that loss-aversion tendencies are overcome. The trade-off between these different issues clearly influences the outcome of household risk diversification strategies.

Secondly, the model is expressed in aggregate terms and it does not explain how potential migrants are chosen from within households, although this is presumably a reflection of human capital differentials and hence potential earnings in different labour markets. It also does not explain how potential migrants choose specific destinations, although this implicitly depends on the distribution of different types and levels of economic risks across different markets, and the realistic potential for households to be able to combine these. There is evidently a need to combine the NELM model with the research in behavioural economics on risk tolerance and competence to manage risk, as well as on heuristics, although the shift from the individual to the household level is conceptually challenging.

Thirdly, the main emphasis in the NELM perspective is on how migration can mediate risk for the household as a whole. However, as indicated in our earlier discussion, the net change in utility need not be positive, because the impact of migration on production constraints is not necessarily positive. The resulting labour shortages to the household caused by migration may result in labour shortages which are an obstacle to investing remittances in higher-return but labour-intensive activities (Taylor and López-Feldman

2007: 3–4). These labour deficits can be overcome if there is a local non-household supply of underutilised labour, but that is contingent on institutional and local economic considerations which are likely to be variable. Therefore, selective outmigration can actually increase the vulnerability of the household members that remain in the home region's economic space. Unless they receive sufficient remittances, and have sufficient labour, to transform the traditional production of the household, the incomes of those who remain behind may decline or at best stagnate. As a result, they become highly dependent on the migrant remittances, and their exposure to risk may actually be increased. The extent of that vulnerability depends not only on the level of risk that the source of income abroad is exposed to, but also to the strength of the social contract amongst the household members (which is discussed below).

The vulnerability of those left behind tends to be age- and gender-specific. Villages full of older people have their own risks, quite apart from their exposure to the risks to the remittance flows from sons or daughters. And, in terms of gender, Rogaly and Rafique (2003) found that some of the wives of seasonally migrant men had to borrow food or ask for other forms of assistance on a persistent basis. De Haan *et al.* (2000) reports a similar situation in Mali, where some women had to rely on food provided by their extended family, and on the labour of their neighbours in order to cultivate their land. Hence, as Hamid (2007) argues, selective migration of household members has both costs and benefits, and constitutes a 'double-edged sword'.

Fourthly, in keeping with its rationalist perspective, the NELM model seems to assume that there is an approach to decision making which evaluates all the available options, and then reaches a unified decision. That decision is assumed to be unproblematic, and probably consensual, and this is a common feature not only of the NELM but of much of the earlier research on family decision making (Blood 1962). In contrast, Shepherd and Woodruff (1988) argued that conflict is as likely as consensus in family decision making. It is not so much a rational process as a muddling along process, where discussion ranges over and around the same issue, with partial and contingent decisions being taken and retaken, and the participants having varying degrees of satisfaction and dissatisfaction with the final outcome. This is, of course, likely to be exacerbated where there are high levels of uncertainty involved.

There has been increasing interest in family conflict management. In one of the earliest examples, Doyle and Hutchinson (1973) identified three main areas of conflict relating to who should make the decision, how the decision should be made and who should implement the decision. This has direct application to the risk-related migration decision within the household, which is one of the most important family decisions. The key questions are who makes the decision about whether a household member should migrate, how should that decision be made—including the degree of consultation with other members—and who should migrate. Underlying this are other issues, such as who will determine the different alternatives to be considered (including national versus international migration) and assess the different levels of risk attached to these. One perspective on the different power structures which lie behind both conflictual and consensual family decision making is provided by Pahl (1990) who, in a non-migration study, identified four patterns of financial control within families: wife-controlled pooling, husband-controlled pooling, husband-controlled or wife controlled. The terminology is deliberately chosen to indicate that even supposedly joint decisions may be based on unequal levels of control between those involved. Similar pooling may occur in relation to both resources and risks in the case of the NELM model.

Fifthly, as was noted earlier, the NELM model is implicitly based on the notion that a social contract exists between the household members—about the migrant sending remittances, and about his/her access to these pooled resources on return. But what guarantee is there that this system of co-insurance will be enacted, and the implicit social contract will be adhered to? Will the migrant be tempted to disown the contract, after securing a job and income at the destination? And what happens if he/she marries someone at the destination, and starts his/her own family? Stark and Lucas (1988) contend that the social contract within the family is underpinned by altruism, but they also recognise that the extent of that altruism is likely to vary across families in relation to the degree of social cohesion and social bonding. In short, 'the new economics of labor migration implicitly assumes a cohesive, traditional family, the members of which share common goals, and are likely to trust, and remain loyal to, each other' (Stark and Lucas 1988: 469). Altruism is also interwoven with self-interest, such as expectations about inheriting land or housing on returning, at the end of the period of migration (Stark and Lucas 1988). In contrast, the migrant who—at some stage—expects not to return home will probably have less reason to adhere to the social contract. This is acknowledged in the concept of 'tempered altruism or enlightened self-interest' (Lucas and Stark 1985: 901). Motivations are, however, gendered—which is not acknowledged in the NELM—and sons are more likely than daughters to be motivated by inheritance prospects, that is, by self-interests (Hoddinott 1992).

6 Constructionist Approaches
Culture, Identities and Risk

INTRODUCTION

So far this book this book has focussed on positivist, or rationalist, approaches to risk, which consider it to be objective, real, 'out there' and measurable. In contrast, constructionism considers risk—like knowledge—to be blurred, indeterminate and socially constructed. In some ways, this can be understood in terms of postmodernist thinking (Hassan 1985). Bauman (1987), for example, argues for perspectives that are less categorical and conclusive than modernist perspectives. This section begins by considering the implications of postmodernism for analysing migration and risk, before discussing constructionism. In the second section we consider the work of the anthropologist Mary Douglas on risk, and consider how this can be applied to migration, before concluding with a review of the role of social identities in relation to migration and risk. We return to constructionist approaches in Chapter Nine in relation to public and policy discourses on migration and risk.

A NOTE ON POSTMODERNISM

Postmodernist approaches emphasise diversity, and that individual migrants are active rather than passive participants. Meanings are not passively accepted but actively negotiated in relation to power: they are constitutive activities (Fraser 1989). Individual migrants deal not in absolute truths about home and the destination, or the migration process, but in relative truths. They understand the world as complex and shifting and, as Denzin (1991: 151) contends, engage in 'both-and' rather than 'either-or statements'. In terms of risk, they are not seen to be dealing with calculable probabilities and rational decision-making processes. Instead, they are involved in compromises and seek outcomes that combine different options, for example, in terms of how visits home, visits from friends and relatives and the duration of migration interact as a series of trade-offs. They may go somewhere that is a 'risky' destination because of antagonism towards immigrants, but decide that this can be managed by living within an ethnic enclave. Migrants' decisions may rely as much on affective or emotional knowledge or intuition as on intellect (Styhre 2004). This resonates with Williams and Soutar's (2009) view that risk has a strong and immediate impact on individuals' emotions, and that these in turn influence their decisions and feelings about the outcomes. Reading about a conational migrant being murdered in New York or London may play out strongly on the potential migrant's emotions, irrespective of the 'objective' levels of risk in these cities.

Migrants have blurred and shifting images of the places they migrate to. These are based on narratives about those places, narratives that are told in newspapers, books, television programmes and the internet, but also narratives that are told by friends, and even by casual acquaintances. Drawing on Shields (1991), Crang (2004) exemplifies this in terms of how 'the northern wilderness of Canada becomes scripted as an arena for masculine endeavour—an environment for adventure, for men especially to prove themselves'. These narratives create, sustain, and re-create myths about the destinations—whether the 'streets are lined with gold', or instead are 'lined with hostility to, or risks for, the migrant'. The myths are not unchanging, but are challenged and rewritten, and are told differently to different audiences. Migrants abroad may discuss with each other the precariousness of their jobs or the risks of being estranged from families at home, but when they do return home they may tell a different story—of their success at work, of their new friends, of their confidence in a relatively certain future.

Of course, migrants do not leave their cultural histories behind them when they set off for the destination. But they are aware of the scripts about risk and behaviour in the destination, and what is expected of them. However, even then there is negotiation between different values, identities and practices relating to risk. They cannot simply forget or compartmentalise the rules they learnt about risk at home. And the migrant and the migrant destination are fluid and mutually shaping: the migrants contribute to changing the scripted rules of risk in the destinations, as well as performing these rules. This process of negotiation is particularly evident in transnationalism. Although the implications for understanding migration and risk have not been explicitly investigated in the literature on transnationalism, this 'is closely linked with broader interests emerging from postmodernist and feminist theory to theorise space and place in new ways' (Brettell 2008: 121).

This process of negotiation continues after the migrant returns to the destination. He or she cannot simply unlearn all the experiences of the migration year, and just pick up their previous risk-taking practices when they arrive home. They will have changed, and the places they return to will have changed, and a whole set of new negotiations of risk will have to take place. This can lead to disorientation, which is particularly marked for the children of migrants, as Mandel (1990) recounts in the case of adolescent Turkish returnees. They may talk about their experiences, and these narratives may influence their own behaviour and that of their listeners. Their migration experiences are part of their cultural capital (Bourdieu 1984), and this influences how they narrate their experiences both of being abroad, and of returning. For example, backpackers produce narratives about their sojourns to particular places, and in part these are told in terms of thrills, adventure and danger, which may either reinforce or challenge the imaginaries of other backpackers about these places (Elsrud 2001; discussed more fully later in this chapter). In short, postmodernism tells us that risk is not 'out there', unchanging and having the same meaning to all migrants. Instead, it is contested, shifting, incorporated into narratives about destinations and home, and transformed by the migrants' own narratives.

RISK AS A CULTURAL AND HISTORICAL CONSTRUCT

How individuals understand and respond to risk is culturally informed. That is implicitly recognised even in the work of the behavioural economists who have demonstrated significant intercultural differences in willingness to take risk, with, for example, this being greater in the USA than in Germany (Fehr *et al.* 2006). There is also a debate as to whether the population of the USA has become more risk averse over time (Box 6.1).

Box 6.1   Are American Workers More Risk Averse and Less Likely to Migrate?

The willingness to take risks, in pursuit of an individual dream, is a long standing part of the American psyche, but some commentators have argued that Americans' risk-taking spirits are waning. In an article in the *Wall Street Journal*, Casselman (2013) argued that there were three distinctive changes which suggested that Americans had 'turned soft on risk': the slow rate at which companies added jobs even in boom times, the reluctance of investors to take a risk on new ventures; and that 'Americans start fewer businesses and are less inclined to change jobs or move for new opportunities'. The latter is of particular interest to this book.

Individual workers traditionally were seen as being willing to undertake small risk-taking acts on a regular basis, quitting their jobs to look for better opportunities, or being willing to move to find work as unemployment increased. The outcome—according to neoclassical economic perspectives—was that 'these acts of faith and ambition help speed money, talent and resources to where they are needed'. Such risk taking contributes to high rates of churn in the labour market, and this flexibility was held to make an important contribution to the effective allocation of resources in the economy, and to high levels of productivity.

Casselman quotes a number of indicators of the decline in risk taking, one of which was the reduction in internal migration in the USA. Data on interstate migration shows that migration rates have fallen over at least two decades. Workers are also less likely to be mobile between employers: the proportion who had been in the same job increased from 46% in 1993 to 53% in 2012. Unsurprisingly, there was a relatively high correlation between occupational and geographical migration. To some extent, the reduced migration was due to the housing market collapse after 2008, and to rising health care costs which have made it riskier to leave jobs, while immigration rules also deterred potential entrepreneurs from entering the USA. An ageing population is also important because young people are more likely to be mobile, but this is not the full answer, because young workers have also been changing jobs less often.

Therefore, while there does seem to have been a decline in risk taking, the causes of this are unclear. Whether and why the culture of risk taking has declined in the USA remain open questions.

Source: Casselman (2013)

Another study (Weber and Hsee 1998) demonstrates that Chinese savers/investors tend to be more risk seeking than American ones. The authors of this study advance the 'cushion hypothesis' as an explanation, that is, individuals will tend to be less risk averse if they can rely on social support to provide a minimalist financial safety net, as in China. This is also associated with China having a collectivistic culture compared to the more individualistic USA. Such cultural differences are the starting point for social constructionist approaches. Taylor-Gooby and Zinn (2005: 5) write that:

> Socio-cultural approaches . . . include a wide range of cultural bases for risk perceptions, all sharing the view that cultural assumptions across social groups are powerful bases for ideas about risk and how to deal with it. They offer an important alternative to the individualistic and rational actor accounts of risk responses developed primarily in economics and psychology. Commentators have suggested that the lens that sees most people's risk perception as shaped by an over-arching culture may direct attention away from the extent to which specific and often local cultures and understandings may offer helpful insights into risk responses, just as it undermines any realism in identifying risk.

In essence, and as Zinn and Taylor-Gooby (2006a: 37) argue in a different publication, this is an approach that contends that we need to move beyond individual cognition, and need to understand individual responses to risk 'against the background of their embeddedness in a sociocultural background and identity as a member of a social group'. Writing specifically about risk perception, Masuda and Garvin (2006: 438) similarly argue that whereas research has tended to view 'individuals as atomised units unconnected to a social system, we now understand risk as embedded in social

---

Box 6.2   Religion and Attitudes to Risk in Pirogue Migration in Senegal

Young men in Senegalese fishing communities have engaged in migration by pirogues (small fishing boats), even though this involved a considerable risk of death. They expressed their attitudes to death through the lens of their religious belief, but this also influenced their attitudes to migration and return. They considered death to be a natural part of life which was, therefore, determined by God: many commented that 'we only die once'. Their religion also taught them that only God, and not other men, was to be feared. These beliefs made them at least appear to be relatively at ease with the possibility that they could die while migrating in these small fishing boats. If that was the outcome, then it was considered to be God's will, their inescapable destiny as dictated by that higher being. This determinism made them appear relatively fearless of the very considerable risk of death. Some parallels can be drawn to the fatalists in Douglas and Wildavsky's (1982) typology.

The risks of death were so high that some community members considered that it effectively represented 'suicide'. However, as this was counter to their religious values, this was roundly rebuffed by the migrants. Instead, pirogue migration, 'framed in a narrative of sacrifice and duty, may provide a sense of closure through a means that is both morally acceptable and symbolically congruent with social expectations of men transitioning to adulthood' (Carretero 2008: 47). Being able to negotiate the risks of such migration was also closely associated with masculinity, and the notion of what it meant to be a man in those communities. Their deterministic view of death did not, however, mean that they were totally passive in the face of death. Rather, they underwent various forms of spiritual preparation before the journey, involving prayers, obtaining amulets, or performing sacrifices—in effect seeking God's help to protect them from the possible negative outcomes of these risks.

Source: Carretero (2008)

---

context'. One important expression of this is the notion of 'the culture of migration', where migration is deeply ingrained in both values and communities in particular contexts. It has become a rite of passage, comparable in many ways to marriage (Cohen 2004). One of the most important determinants of attitudes to risk is religion, as Carretero (2008) reports in a study of Senegalese fishing communities (Box 6.2). These are communities where the culture of migration is strong and persistent, even in the face of risk of death.

One example of the importance of shifts in the sociocultural background is the way in which the West in the USA changed from being considered 'wilderness' by the early migrants from the east coast, to being seen as a highly attractive and desirable environment by the twentieth century. Similar accounts can be provided in Europe; for example, how from the 1960s the Spanish Mediterranean coasts shifted from being seen as underdeveloped, arid and difficult environments, to being attractive to lifestyle migrants from Northern Europe (King *et al.* 2000). There are multiple reasons for this, including structural demographic and economic shifts, and investments in infrastructure, but there have also been sociocultural changes in terms of how risk is perceived, and the narratives about risk or absence of risk in these areas. As Tulloch and Lupton (2003) comment: 'Risk knowledges, therefore, are historical and local. . . . As a result, risk knowledges are constantly contested and are subject to disputes and debates over their nature, their control and whom is to blame for their creation' (Tulloch and Lupton 2003: 1).

While some researchers do not believe that the cultural and the rationalist individual approaches can be combined, others contend that there is scope for this. A notable example is Kasperson *et al.*'s (1988) SARF approach, which has been applied to explain the ways in which social context can influence communication about risk events (Pidgeon *et al.* 2003). The SARF model contends that social and institutional facts influence risk perceptions via socially mediated communication channels such as the media and public relations campaigns, as well as via social networks. Examples could involve campaigns to influence people to migrate or to deter potential migrants, and there are cultural differences in such risk communication (Kasperson 1992).

More generally, risk in any specific social setting is not determined objectively: 'in real life situations, the boundary between certitude and uncertainty is of course seldom razor-sharp, and vagueness and ambiguity tend to be the rule rather than the exception' (Boholm 2003a: 168). Boholm (2003b) concludes that we need to consider 'situated risk', and that anthropological and ethnographic methodologies allow risk to be contextualised. This means looking at local and national value systems, practices of risk management, and the socioeconomic and sociocultural factors that influence how individuals respond to risk. One aspect of this is the importance of the social groups that individuals belong to, because risks should be understood in the context of the interactions of individuals, groups and institutions (Masuda and Garvin 2006). This is an issue which also particularly interested the anthropologist Mary Douglas, especially why some cultures strongly focus on risks that other cultures see as being of relatively limited significance. The answer to this question is complex, but one of the most important aspects of situational risk is how individuals are positioned in relation to social groups. This is considered in the next section.

MARY DOUGLAS AND THE ANTHROPOLOGY OF RISK

Anthropologist Mary Douglas's ethnographic studies in Zaire challenged what she saw as a prevailing objective and technological conceptualisation of risk (Zinn 2004a). For Douglas, risk had 'a social construction in a particular historical and cultural context', so that it was open to varying social interpretations (Douglas and Wildavsky 1982: 6–7), that is, she saw risk through a constructionist lens. Societies fear different types of threats, and these are related to their ways of life and cultural biases. Her research examines how a traditional moralism of pollution, which had evolved in relation to the understanding of sin within an essentially religious framework, was transformed over time to a more secular approach which understood threats as risks (Taylor-Gooby and Zinn 2005). One of the central tenets of her findings was that 'blame' can be associated with either the victim, the individual at risk, or to the perceived cause of risk, typically seen as 'blaming the outsider' (Douglas 1985: 59).

There is a long tradition in anthropology of using 'typologies as a way to theorize about similarity and difference' (Brettell 2008: 115), and Douglas's work is no exception in this respect. Douglas (1992) proposed a typology of risk rationalities based on notions of social organisation as captured by the terms 'grid' and 'group'. These formed the principal axes of her two-dimensional, four-fold typology, which was set out in detail in her coauthored book, *Risk and Culture* (Douglas and Wildavsky 1982). It is a typology which reflected the importance attached to insiders versus outsiders, as indicated above. 'Grid' essentially described the extent to which norms structure action, and individual decisions are constrained by their societal position. In contrast, the notion of 'group' indicates the extent of group cohesion and the boundaries with the external world, and with outsiders versus insiders. In essence, group is about the extent of the solidarity within a particular society. The typology has been one of the more influential contributions to research on how risk is understood. It proposes four major biases: Individualist, Egalitarian, Hierarchist, and Fatalist, plus a fifth asocial Autonomous perspective which was added in a later work by Thompson *et al.* (1990).

The individualist is low on grid and low on group. He or she is relatively unconstrained by society and also lacks close social ties. (S)he tends to be enterprising and self-made, willing to experiment, and seeking to control the environment (s)he operates in, and the people they operate alongside of. The individualist is also highly resilient in the face of risk and external shocks, and in part this is because of having a strong belief that the system has capacity to recover from such shocks and negative impacts. The outcomes of his or her activities are assessed in terms of the wealth accumulated, and the number of people that control is exerted over.

The hierarchist is high on grid and high on group. He or she sees the world as having strong group boundaries, and being strong on constraints. There are well-defined roles to be occupied in a world with clear and generally recognised rules, as typified by India's caste system. The hierarchist believes that it is possible to exploit the world around you, but only within limits, because to exceed these is to incur the risk of collapsing the system.

The egalitarian is low on grid and high on group valuing solidarity and the notion that being egalitarian stems from placing the good of the larger group above that of the individual. The egalitarians are also sensitive to risks to nature and the environment, even if these have low probabilities. They have strong adherence to traditional ways of life and are highly conscious of risks to these ways. They generally have low respect for externally imposed rules.

The *fatalist* is high on grid and low on group. Fatalists consider that they have relatively limited control over their own lives in the face of externally imposed constraints: they do not belong to those groups which have power to shape their way of life. Unable to manage effectively their situation, they tend to be passive and to resign themselves to their fates.

The *autonomous* individual tends to be withdrawn from the surrounding society and has minimal interactions within it.

Each of these idealised types arguably can be associated with different migration risk rationalities, although this is probably clearest in the case of the individualist and the fatalist. They do not predict whether individuals will or will not migrate, but they do provide insights into the rationales behind these decisions.

Individualists favour a market culture and would explicitly be expected to be risk-takers with respect to migration, given that they are governed more by individualism, and have relatively few social ties. They are enterprising and seek to take control of their lives, and if migration is the means to achieve this, they are more willing to take the associated risks. Individualists are typified by the early migrants to new destinations, those who believe that although this may severely disrupt their lives, they nevertheless will be able to succeed through their own initiatives. But individualists may also decide not to migrate, believing that they can make their way in the world in their home locality.

In contrast, fatalism is characterised by passiveness in the face of being unable to control one's life. This accords with the notion of divine determinism which, although present in all societies to some degree, is particularly strong in some. It may contribute to individuals deciding not to migrate even in the face of considerable negative risks, because they believe that little can be done about their preordained fates. Or, as in the example of the Senegalese fishermen (Box 6.2), the migrants accept the high risk of death because it is understood to be bearable as it is God's will whether or not they will die during the course of their dangerous voyage.

The hierarchical type is characterised by strong social cohesion, and low freedom of movement, because individuals have well-defined roles in a world with strong rules. Hierarchists believe they can take risks, but only within tight limits. In terms of migration, we can speculate that they may be willing to participate in group migration, along with other members of their household or community, where this is informed by strong social cohesion. The egalitarian or enclave type is characterised by strong social cohesion and a high degree of freedom to take risks. Egalitarians value traditional ways of life and this may make them reluctant to migrate. But they also place the good of the group above their own welfare, and this could accord with the notion of the individual who migrates in order to diversify the sources of household income, even though they may face considerable personal risk as a consequence (Chapter Five). But, equally, they may be willing to sacrifice their wish to migrate, in order to provide support for those who would be left behind, perhaps elderly parents. If autonomous individuals do migrate, they are likely to be loners, and to shy away from ethnic enclave communities at the destinations.

This typology does provide useful insights by presenting four ideal types. However, as indicated, it provides insights into rationales relating to risk rather than some simple association between each type and the propensity for migration versus non-migration, or particular types of migration. There is clearly potential to investigate whether the typology can be applied to empirical research on migration, but a note of caution is required because the idealised types are 'too schematic to grasp the complexity of social life' (Zinn and Taylor-Gooby 2006a: 39).

Another important issue relates to causality—are individuals fatalist because of their experiences of having little control over their lives, or do they believe they lack such control because they are fatalists (Adams 1995:64)? In one of the few studies of this issue, Dake (1991) analysed whether individual concerns about a range of issues could be predicted from the four-fold grid-group typology. Although there were some statistical associations, the study was not able to determine the direction of causality.

Douglas's work, despite the difficulties indicated above, has been influential in the field of cultural theories and specifically in relation to risk (Adams 2001). The notion of culturally constructed risk has been linked to myths about nature by Schwartz and Thompson (1990). Their particular concerns have been to understand the origins of the beliefs about nature that influence decision making under conditions of risk. By extension, the same approach could be applied in migration studies—whether in respect of the migration process, or of the migrants themselves. There is research on the culturally constructed othering of migrants by, for example, Burchardt (2004) and McLaren and Johnson (2004), and we return to this theme in Chapter Nine, but the links to Douglas's work have not been empirically explored in migration studies.

SOCIAL IDENTITIES AND RESPONSES TO RISK

Douglas's typology and ideal types have, inevitably, been criticised for over-simplifying the complexities of the relationships between risk and culture, leading to calls for a more multilayered approach (Zinn and Taylor-Gooby 2006a: 39).

One strand in this has been the critique of Douglas's work as being too static (Bellaby 1990). Cultures are living and dynamic, and may be shifting between, or combining, elements that are associated with the grid-group typology. This means that it is difficult to allocate individuals, or the social groups they belong to, unambiguously to one of the four ideal types. Individuals over the course of a lifetime may reflect a high degree of flexibility and change. They may change their identification with a particular group, or the constitutions of these groups may change. The acceptability of risk changes over time so that 'the boundaries or limits which determine risk behaviour differ for different people' (Rhodes 1997: 220).

An important extension has been to move away from ideal types, and instead to examine social identities and membership of cultures and subcultures. This is an important counter to, say, the work of the behavioural economists, who focus on individual risk tolerance and risk aversion. In contrast, the cultural approach emphasises the role of the group that you belong to, or identify with. Even individuals with similar levels of risk tolerance may act differently because of different levels of risk tolerance in the groups they identify with (Tulloch and Lupton 2003: 1). This is consistent with Green's (1997) argument that individuals construct social boundaries through the narratives they present, or exchange with others, about risk; they forge subjective social identities. Migrants or potential migrants could similarly be expected to be influenced by the risk tolerance prevalent in the groups they identify with (Heitmueller 2005). However, these need to be seen as changeable and flexible.

Perhaps of more direct relevance for migration is the way in which Barth (1969) emphasises ethnic identity as being fluid and contingent. Gupta and Ferguson (1997) similarly consider that there are multiple and shifting bases to self-representation and identification. Identity is not fixed, and neither is there an identity within you that is waiting to be discovered when engaging in a risky activity such as migration. Instead, identity is constantly being revised, has many aspects, and is presented differently to different people. It can be seen in terms of the 'life story' whereby individuals order events in their lives, partly as a way of explaining the formation of their identities (Alheit 1994). However, these biographies are also full of discrepancies, inconsistencies, and tangential wanderings that are woven into, and rewritten, as part of this constantly changing life story (Ochs and Capps 1996).

Brettell (2008: 132) takes us further because she explicitly addresses issues of identity in relation to migration: 'The act of migration brings populations of different backgrounds into contact with each one another, and hence creates boundaries. It is the negotiation across such boundaries, themselves shifting, that is at the heart of ethnicity and the construction of migrant identities'. Backpacking is one relatively well-documented type of temporary migration, bordering on long-stay tourism and travel, which demonstrates this process of identity formation and negotiation. Box 6.3 sets out an illustration of identity formation amongst backpackers, emphasising the narratives they present differently to different audiences, as part of projecting contrasting images of their identities. This can be seen in the terms of Sharland (2006: 36), addressing the issue of risk taking amongst young people more generally. She contends that:

> We need to look more closely at what risk taking means to young people, its dynamics, and the relationships and resources surrounding it. We might also draw on some particular concepts, risk culture, cultural learning, identity capital or habitus to elucidate. Most importantly, we must recognise that risk taking is integrally bound up with the development of young people's identities. To problematise this is a necessary and productive activity for informing practice. But to consider risk taking always and necessarily problematic would be missing the point. We must start by recognising risk taking as a routine, even desirable, component of young people's lives and development. Where and how we begin to define it as troubling or troublesome must then be up for scrutiny.

Box 6.3   Backpackers, Identities and Risk

'Risk and adventure narratives' are common amongst backpackers, and contribute to the way in which independent travellers project their identities as being able to take and to manage risks. They are highly gendered, they tend to stereotype remoter places (in the Third World) and are constantly shifting (Dake 1992). 'With such a constructivistic approach, risks are sometimes nothing more than social or cultural constructions with the conscious or unconscious purpose to maintain a cultural structure' (Elsrud 2001: 598). At the same time, the backpacker is engaged in a 'self-reflexive project, and that is worked out by the absorption, creation and recreation of myths'. The myths are produced both by the backpackers, and by and in various forms of media.

Perhaps the strongest backpacker myth is about how the individual, or a small group of friends, have encountered—meaning risked encounters with—remote, 'untouched' areas and people. They have travelled long distances, mostly on local transport, or walking or cycling, and have lived in remote communities. They have also eaten local foods and observed, and perhaps participated in, local ways of life and cultural practices. In Giddens's (1991: 124) terms, they are actively courting risk, while Goffman (1967: 260) emphasises that they are undertaking activities 'outside the normal round, avoidable if one chose, and full of dramatic risk and opportunity'. Their 'strong character' in the face of such risks (what behavioural economists term 'competence' to manage risks) is subsequently incorporated into the narrative they tell, which establish their identities as independent, resourceful and risk taking (Riley 1988). But it is not the content of such acts which define them as being with or without risk: 'It is how the act is experienced, when and where it takes place, and what mythology has to say about it that creates the definition' (Elsrud 2001: 603).

The backpacker does not only relate these narratives to other backpackers, but the narratives also travel home with him or her. They gain significance because 'the very progress of many Western societies is often related to risk taking. Progress, it is said, relies on people and society investing extra while hoping to gain even more, on individual effort expended in order to reach personal and communal gain, on moving ahead by daring to face novelty' (Elsrud 2001: 613). Risk taking through mobility plays directly to this belief.

Source: Based on Elsrud (2001)

Sharland encourages us to see risk attitudes as dynamic, but also as inextricably bound up with the creation and sustaining of identities. The managing of risk by young migrants and backpackers is also inextricably bound up with identity formation.

CONCLUSIONS

This chapter constitutes an important turning point in the book, when our focus shifts from understanding risk and uncertainty as objective and 'out there', to being subjective and socially constructed. It is an argument that the meaning of risk and uncertainty are highly contextual, depending on how actors or events are socially situated, and recognizing the long historical trails that have shaped understandings and practices. There are competing imaginaries of risk and these do not arise in a social vacuum but are defined by practices and interests, past and present. There are no fixed points in this theoretical view, but a series of often blurred and shifting understandings.

While this perspective is particularly important in encouraging a deeper, more critical and reflective understanding of risk, it can often be difficult to apply in empirical research. The generic work of the anthropologist Douglas on risk provides one approach—based on a typology of social groups characteristics—to construct a theoretical framework that provides a structured approach to understanding migration, risk and uncertainty, although most postmodernist and cultural theorists, nowadays, would see it as potentially too inflexible and inherently a static approach, given the constant changes in the nature of social group formation and practices. Hence, for many commentators, social identities provide a potentially more fruitful approach to the subject. We return to further consideration of constructionist approaches in Chapter Nine in relation to public and policy discourses on migration and risk.

7 The Role of Networks and Intermediaries in Mediating Risk

INTRODUCTION

We conclude the analysis of the individual and household levels with a discussion of how individuals mediate risk through using intermediaries. There are several theoretical perspectives on this, some of which have been touched on in earlier chapters. For example, when considering the rationalist approach of the behavioural economists (the notion of 'source preference', Fox and Tversky 1995: 600), or what cultural theories tell us about how social identities influence key individuals or group decisions. This chapter shifts from the micro to the meso level, and to the consideration of how both formal and informal intermediaries mediate the relationship between risk and migration.

First we consider the role of informal networks, typically constituted of friends, family and acquaintances, including those who are encountered in digital spaces. One advantage of informal channels is that they can supply potential migrants with tacit knowledge, which is difficult to acquire by other means (see Williams and Baláž 2008, chapter two). Migrants also find it easier to gather information from friends and family, compared to official information channels, and to secure information on difficult-to-codify elements such as the emotions experienced, whether the disappointments or the joys of migration. Moreover, it is likely that potential migrants do not even know what all the important factors are—that is, areas of knowledge they require—about the destination, until they have acquired this tacit knowledge from friends and family. Therefore, the migrant's informal networks can both clarify areas of uncertainty and contribute to forming more realistic estimates of risk. The growth of digital communication and social media has vastly extended the range of friends, acquaintances and strangers that can provide knowledge for migrants.

Secondly, migrants can also utilise formal intermediaries. The most obvious of these is reliance on insurance in the face of risk. The reliance on insurance—a means of collectivising risk—is quite common in a number of arenas of modern life, at least in more developed economies. Other than in respect of health, most migrants make very modest use of insurance policies.

Another form of intermediation used by migrants is migration agencies, which help to reduce the transaction costs and risks of migration. The nature of such agencies has changed over time, due to the growth of digital communication. The agents of individual employers recruiting face-to-face in countries of origin, as in the age of mass migration, have been replaced by specialised intermediaries, by online agencies, and by head hunters for highly skilled workers. Such agencies may mediate but they do not, of course, eliminate the risks and uncertainties faced by migrants. Finally, migrants faced by significant border controls may rely on intermediaries to circumvent these, whether via regular or irregular channels. Typically this contributes to smuggling and trafficking—two very different ways of engaging with risk.

Intermediaries (other than in the forced mobilities of trafficking) have two important characteristics. First, potentially they can reduce the transaction (total) costs of migration because of their superior codified and tacit knowledge of both the destination and the migration process. This ranges from dealing with the sometimes bureaucratic mazes of visas and migration regulations, to networks of contacts with employers, to superior language knowledge. They cannot eliminate either risk or uncertainty, but they can realign the boundary between knowledge and uncertainty.

Second, faced with uncertainty, migrants seek reassurances about the outcomes of their migrations. Therefore, one of the qualities they seek from intermediaries is that they can trust them. In very different ways this applies both to regular migrant labour agencies, and to smugglers. Another classic example is the use of informal value transfer systems to transfer remittances (Box 7.1). In migration, as in many other arenas of life, 'trust begins where knowledge ends' (Lewis and Weigert 1985). There are, of course, important differences in how trust is theorized. Whereas economists conceptualise trust in terms of reductions in transaction costs, sociologists consider trust as given in advance and developed from shared values and routines.

Box 7.1   Informal Value Transfer Systems: Redistributing Risks

There are millions of migrants working away from their home countries who wish to send money (remittances) to their families. Sending money is not without risks, especially when migrants work irregularly. They also may wish to avoid arousing the suspicions of the tax authorities in both the countries of origin and destination. Finally, in some countries and regions, networks of formal banking are underdeveloped or absent. Where formal banking fails, the informal value transfer systems (IVTS) come into play.

The IVTS predate formal banking systems and are established in many different countries and regions: *hawala* in the Middle East, Afghanistan and India; *hundi* in India; *al-barakaat* in Somalia; *phoe kuan* in Thailand or *fei-ch'ien* (flying money) in China. Each type of IVTS has its own specific procedures for transferring remittances, but their basic structures are similar. A migrant passes money to a local IVTS agent in country A. This agent sends notification of the money received to a member of the IVTS in country B via fax, email and/or phone call. The agent in country B delivers money to the migrant's family in a few hours or days. This is done with no formal recording of the international transfers. The IVTS also operate as clearing houses. Settlement between the IVTS members is done either via formal bank transfers or via under-invoicing for goods and services exported abroad (Schaffer 2008).

The IVTS operate with surprising efficiency. For small transfers (usually less than $200), the *hawala's* flat fee, for example, can be 5–10 times less than that charged by the bank. The transfer usually takes less time to realise and is also able to deliver funds to recipients in rural areas, where no formal banking operates. The *hawala* transfers also avoid formal controls on currency flows, such as fixed exchange rates and tax regulations. The IVTS networks are founded on trust, shared beliefs and cultural norms, and dense networks of informal and formal business ties between the IVTS members. The IVTS plays a significant role in international transfers. While officially recorded international migrant remittances were estimated to be $483 billion in 2011 (of which $351 billion flowed to developing countries), unrecorded flows via the IVTS could be at least 50% higher according to estimates by the International Monetary Fund (Ratha 2011).

The IVTS are not unproblematic. The terrorist attacks on 11 September 2001 turned attention to 'underground banking networks' (McCusker 2005). The IVTS (*hawala* and *al-barakaat* in particular) were considered to be associated with tax evasion, money laundering and payment channels for international terrorism. The IVTS probably were used in money transfers to terrorist groups, but these groups also used the regular banking system, and migrant remittance networks were often 'needlessly criminalised' as risks to society (de Goede 2003: 528).

Sources: Ratha (2011); Schaffer (2008); McCusker (2005); de Goede (2003)

NETWORKS: KNOWLEDGE TRANSFER AND RISK

Networks are a resource that can mediate risk. In this section, we consider their role in mediating decisions about where to migrate to. However, because migration may also disrupt existing networks, leading to a decline in the resources available to migrants, it can also increase the risks they are exposed to. Social capital is not easily transferred between places, because it tends to be location specific, so that migration is often associated with loss of social capital (Fischer *et al.* 1997). Therefore, one factor that has to be considered by potential migrants is the risk of losing existing close ties and social capital, and the uncertainty attached to building new networks in the destination.

In a more positive vein, informal networks of families and friends can mediate risk, by providing information, housing support and assistance to find jobs. Haug (2008: 588) defines such migration networks as 'a composite of interpersonal relations in which migrants interact with their family or friends'. The networks are likely to include both those with and without migration experiences, who can provide knowledge about the risks attached to different options in both the home region and potential destinations. Not all friends and relatives are likely to be equally trusted sources of knowledge. If you are intending to migrate to, say, London, you are more likely to trust the knowledge possessed by those who have lived there, or who have at least visited. But there may also be important distinctions made between best, close and casual friends. Close friends—unlike casual friends—are especially likely to be trusted. As Annis (1987: 354) comments, as a 'friendship develops, an intricate web of reciprocal and mutual dispositions, beliefs, understandings, feelings, etc., develops'. As these intensify, they provide a basis for trust, founded on a belief that the close friend shares your dispositions and feelings about migration. Close friends are also defined in terms of being willing to participate in self-disclosure (Jourard 1971), and this type of confidential disclosure of inner feelings can be critical in helping to clarify attitudes to the risks associated with migration.

There are a number of ways in which networks—which embrace the origin, potential destination and other places—can influence migration. In one of the deeper studies of this topic, Haug (2008: 588) identifies five main forms of influence:

1. Affinity: strong ties to relatives and friends in the region of origin tend to reduce migration.
2. Information: relatives and friends at the destination act as information channels, thereby increasing the propensity to migrate (Coombs 1978).
3. Facilitation: Friends and relatives tend to funnel migration to particular locations because they reduce the costs of migration to these destinations, through provision of support for finding jobs and houses (Choldin 1973).
4. Conflict: conflicts within your network, in the family or community, may encourage migration.
5. Encouragement: individuals may encourage other family members to migrate for a variety of reasons, including the diversification of household income in the face of risk (see Chapter Six).

Massey *et al.* (1993) particularly focus on the facilitating role of networks in mediating risk. They emphasise that well-developed migrant networks play key roles in helping migrants secure jobs, and therefore reduce the risks attached to future income, especially unemployment. Informal networks also provide support in terms of funding migration, providing accommodation on arrival and crossing borders, whether legally or illegally (Böcker 1994). Social networks provide a means for disseminating information, assistance and patronage, with the latter indicating the implicit power relationships that are embedded in these networks (Haug 2008: 588–589), and which, as we discuss later, carry their own risks. Economists conceptualise the advantages of networks in terms of network externalities (Chiswick and Miller 1996; Church and King 1993; Epstein 2008).

There are diverse sources of knowledge that migrants can draw on including books, websites, and newspapers, but arguably the most valued resources are their social networks (Gamburd 2000), constituting location-specific capital (Da Vanzo 1981) which can lower the costs of migration (Massey *et al*. 1993). Roberts and Morris (2003: 1256), drawing on Fafchamps (1992), elegantly express this:

> The advantages of networks are thus the same as those of households—through interpersonal relationships, individuals acquire "relation-specific" information within social networks that reduces monitoring and transaction costs, allowing these larger units to more effectively insure against risks.

Roberts and Morris (2003) consider whether this argument can be extended. They review the possibility that membership of a network will both reduce the risks of job finding, and increase the range of jobs available to the migrant. The former represents a form of insurance, and the latter increases opportunities and the potential to secure a higher income—that is, it reduces the risk attached to a specific income target. If these relationships hold, then it would be expected that the value of the network to the migrant would increase with its size and diversity. However, there is little empirical evidence for either of these relationships in research on Mexico and the USA by Espinosa and Massey (1999), although that particular study used remittances as a measure of insurance/risk, which is problematic because of the other influences which determine these, including altruism (see Chapter Six). While the evidence of an insurance effect is questionable, social networks are an important determinant of whether to migrate, and the choice of destination (Boyd 1989; Faist 1997; Fawcett 1989).

In behavioural economics, the concept of source preference also provides an insight into the role of networks in mediating risk and uncertainty. Source preference refers to favouring one source of information over others. Potential migrants potentially can collect information on migration destinations from various resources (media, internet, official information provided by destination country, etc.). Much information, however, is tacit in nature and this includes some of the more important aspects, such as how best to fill in visa applications, the costs of living at different stages of migration, or the realistic chances of getting particular jobs. Members of ethnic groups and/ or family networks tend to be considered the most effective sources of tacit knowledge, as they are characterised by higher levels of trust and lower communication barriers compared to other carriers. Information and support provided by social networks is also important in transforming severe uncertainty to computable risk and/or certainty about the outcomes of migration (source sensitivity). Member of family or kin, for example, may provide a 'sure job' for the first-time migrant. For a migrant, turning possibility to near certainty via social networks may be more important than increasing probability via a more extensive information search.

Networks which connect migrants, former migrants and non-migrants in the origin and destination regions also contribute to circular and chain migration (Boyd 1989). This relates particularly to the information and facilitating functions identified by Haug (2008), mentioned above. After migration by the first individual from a particular generating region to a destination, the monetary and psychological costs of migration to that place decrease for members of the migrant's social network, due to reduced uncertainty in the decision-making process. There is an interlocking circle here of cause and effect. Reduced costs encourage more migration, while the addition of these new migrants to the migrant network serves to expand it, further reducing risks for future migrants. Ultimately, the risks may be reduced to near zero, simply because the existence of a very well-established migrant community at the destination ensures that the potential migrant has almost as much information about that place as about his/her home region. The nature of the networks also changes over time. For example, there is evidence that family and community networks are substitutes in Mexico-USA migration. Initially, potential migrants rely relatively more on family networks, but as migration to a particular destination becomes established and substantial, there is a tendency to rely more on wider community networks (Winters *et al.* 2001). Networks are also created and re-created by migrants at different stages in the migration process. Umblijs (2012) provides some direct evidence on the effects of network size on the probability of migration (see Box 7.2).

Different types of migrants use different migration channels. Highly skilled migrants tend to be strongly reliant on both formal networks—such as intracompany transfers and academic exchange schemes—and informal ones (Lowell and Findlay 2002; Meyer 2001; Vertovec 2002; Benson-Rea and Rawlinson 2003). With respect to informal networks, they are less likely than unskilled workers to rely only, or mainly, on kin because they have "more extensive and diverse networks consisting of colleagues, fellow alumni, and relatives" (Meyer 2001: 94). In a comparative study of well-educated migrants, Williams and Baláž *et al.* (2005) surveyed Slovak au pair, student and professional migrants who had returned from the UK. All three groups assigned considerable importance to networks of friends, relatives and acquaintances in Slovakia and abroad as sources of information about foreign jobs. However, there were significant differences in the types of networks they used to realise their migrations, not least because they had entered the UK under different visa or work permit regulations. Intracompany transfers and EU-supported mobility programmes had been most important for professionals in general and for scientists, university teachers and other public sector employees in particular. University students were also dependent on

Box 7.2   Migrant Networks and Risk Aversion

It is widely assumed in the migration literature that migrant networks decrease the costs of, and uncertainty about, the outcomes of migration (Massey *et al.* 1993). The immigrant community can provide a new entrant with housing, food and financial support before they find jobs and secure sources of income. Logically, it can be argued that the larger the migrant network, the higher the probability of finding jobs/incomes and the lower the risks of migration. It would therefore follow that a large immigration community would be likely to attract relatively more risk-averse migrants compared to a small community.

There is relatively little evidence to support these assumptions, but Janis Umblijs used the German Socio-Economic Panel Data (SOEP) to analyse the association between the size of expatriate networks and the level of risk aversion for foreign-born individuals arriving in Germany between 1960 and 2000. The total sample was over 50,000, of which some 7,500 were foreign-born. The SOEP contains several questions on risk preferences in various domains (general risk taking, financial risk taking, occupational risk taking, etc.). Field experiments with real payoffs indicated that the general self-assessed risk measure proved to be the best predictor of an individual's actual general risk-taking behaviour. The general risk-aversion measure was rescaled to a scale ranging from 0 to 100, with 0 being the most risk averse.

Umblijs found that the point estimate on the network variable was about –0.05. This means that 'an increase in the network (community) size by 10,000 individuals in Germany reduces the average willingness to take risks amongst migrants by 0.05 points on the risk measure scale' (Umblijs 2012: 20). The five percent difference may seem small, but it is equivalent to the difference between average willingness to take risks by males and females, or the effect of completing secondary school. There were over seven million immigrants in Germany in 2009. So even a small increase in the size of the immigrant community would attract more risk averse potential immigrants, and increase immigration flows. Migrants who arrived in Germany in time periods when their ethnic immigrant communities were relatively large tended to be more risk averse.

As large immigrant communities favour the more risk averse, Umblijs suggested that where national states wanted to reduce inflows of migrants, their migration policies should target diversity and the creation of a larger number of small migrant networks over smaller numbers of large migration networks.

Source: Umblijs (2012)

EU and British Council mobility programmes, not least because of the high living costs and fees associated with studying in the UK outside of such programmes. Language students, on the other hand, had used student placement agencies to find short-term courses. Au pairs, almost exclusively, had relied on au pair agencies, which arranged their visas and other formalities. These networks, in different ways, served to reduce uncertainty.

Reliance on networks is related to the prospective migrants' perceived competence in this arena. McKenzie and Rapoport (2010), for example, studied the impact of education on the use of migration networks by Mexican immigrants in the USA. The probability of migration increased with the level of education in those communities which had limited migrant networks, but it decreased with education level in communities with extensive migrant networks. Perceived competence can also impact decisions to move or stay, and reassess the importance of the network for individual migrants. De Jong *et al.* (1983), for example, showed that rural Filipinos, who had visited Manila several times and/or had migration experiences, were more likely to engage in rural-urban migration. Arguably this reflects the acquisition of mobility competences that make them believe that they can manage the risks of, and achieve higher levels of expected utility from, migration.

A study of the future migration intentions of Slovak returnees from the UK (Baláž *et al.* 2004) also pointed to important changes in the structure of individual migrants' networks after their first migration. More than half the total sample had maintained contacts with friends and colleagues in the UK subsequently, and these networks were considered potentially important for those considering permanent migration in future. There were significant increases in self-reliance after the first move, which mediated uncertainty in finding jobs when considering permanent migration. This was associated with increased self-confidence, as well as the experience of having lived abroad (Williams and Baláž 2005). This raises questions about causality that we have highlighted earlier in this volume: are migrants more risk tolerant because of increased perceived competence acquired through migration, or is migration a selective process, favouring those who were already more risk tolerant?

In contrast to the emphasis on tacit knowledge and learning, Epstein (2008) draws on behavioural economics to provide additional insights into the role of networks in migration. The existence of networks—and the network externalities associated with these (Chiswick and Miller 1996; Church and King 1993)—can identify the range of alternative destinations that may be considered by migrants, but not the reason for specifically selecting one of these. Epstein contends that 'herd effects' provide insights into this choice. In essence, herd behaviour involves discounting the information you have acquired in order to follow the lead provided by others: herd behaviour implies 'I will go where I have observed others go, because all those who went before me cannot be wrong, even though I would have chosen to go elsewhere' (Epstein 2008: 568).

Herd effects is the popular name for the theory of information cascades (Banerjee 1992; Bikhchandani *et al.* 1992). Essentially, it contends that when individuals are faced with uncertainty—having to make decisions based on imperfect information about destinations—they may rely on the decision-making heuristic of imitating the behaviour of similar people: surely if so many other people have migrated somewhere, then not all of their decisions can be wrongly informed. In terms of causality, Epstein contends that networks do not directly generate herd behaviour. On the contrary, information cascades, or herd behaviour, can be based on relatively small numbers of migrants initially, but subsequently they result in large-scale migration to a particular destination that, in due course, creates positive network externalities. In short, 'emigration decisions are made generally under conditions of uncertainty. In such cases, we can only look to herd effects to explain initial immigrant clustering' (Epstein 2008: 580). There is empirical evidence for the existence of both herd behaviour and of network externalities (Bauer *et al.* 2007), but there is not yet definitive—or even very much— empirical evidence on either the direction of causality, or the sequencing of network versus herding effects.

The role of networks has been mediated by technological developments. From the late 1960s, a number of commentators have observed that new technologies were transforming the roles of family and friends (Litwak and Szelenyi 1969). This included both the communication potential of IT services, as well as air travel technologies, which—as with low cost airlines from the 1990s—facilitated more regular visits to friends and kin (Williams and Baláž 2009). Both have served to change the form and costs of interaction with members of networks, both face-to-face and at a distance, as well as the relationship between these. This opened the way to changes in the acquisition of knowledge relating to the risks and outcomes of migration: it became easier to communicate with network members located at the destination, and indeed elsewhere, and to visit them at the destination, thereby acquiring firsthand tacit knowledge. Low-cost airlines also create new geographies of mobilities, and knowledge flows, by creating new inter-regional connectivities (Graham and Shaw 2008). Burrell (2011) argues that the shuttle flights of a company such as Ryanair define the new migrations, and while this is mainly related to the total costs of migration, risk is implicit. The importance of enhanced visits to friends and relatives is summarised by Urry (2012), who emphasises that tacit knowledge and trust still tend to be dependent on at least intermittent copresence.

Ethnic Enclaves, Network Externalities and Risk

There are also considerable differences in how risks are mediated according to whether migrants are acting 'independently' in the destination or are drawn into established migrant communities. As noted above, such enclaves can produce positive network externalities which mediate many of the risks associated with migration, such as finding a job, accommodation, or familiar cultural practices. They also tend to reduce the risk of experiencing discrimination (Grönqvist 2006: 370). Specifically, in terms of risk, Haug (2008: 591) argues that:

> With each new migrant, the social capital at the place of destination increases for the potential successors. In the course of the migration process, the migration risk thus diminishes. Social capital declines at the place of origin, resulting in an attendant drop in the potential loss of social capital at the place of origin. Each emigrant increases the location specific social capital at the place of destination and this accumulation of location specific social capital at the place of destination reduces the opportunity costs of migration for successors. Additionally, staying at the place of destination becomes more attractive as a result of the rising social capital in kinship networks and the ethnic community.

The negative externalities of networks have received less attention than the positive ones in the migration literature. The negative risks include the possibility that while ethnic businesses provide an apparently ready source of initial employment (Portes 1995; Waldinger *et al.* 1990), there is also the possibility of being exploited in dependent relationships in such businesses. This is especially the case where the employer has perhaps sponsored the migration and/ or also provides housing. Migrants may also become locked into relatively closed social networks, which may truncate their networking and their knowledge and experience of the host society. This may result in relative social and economic isolation, while hindering learning the language of the destination, and acquiring language capital (Massey and Denton 1989). This may result in restricted occupational and social mobility, because first jobs can be stepping stones or culs-de-sac in the labour market (Williams 2008). Other disadvantages include the concentration of migrants, leading to intense competition for jobs and the lowering of the wages paid to migrants (Stark 1991).

Expressed in more formal economic terms, there are negative as well as positive network externalities. In addition to the negative externalities already noted, these can 'be subject to diseconomies of size [of the immigrant community] in the immigrant population' (Epstein 2008: 580). Beyond a certain size, there is the risk that relationships become more impersonal, so that newly arrived migrants are less likely to benefit from feelings of either obligation or altruism. There is, therefore, a tipping point at which migrants would find it more beneficial to join a different network. However, herd behaviour may mean that migrants continue to be attracted to such migrant communities, long after the tipping point should have been passed. While the existence of such externalities is difficult to refute, as well as conceptualise, there is a lack of empirical evidence about the relationship between the scale of migration concentration and the extent of the externalities produced.

Haug (2008) provides one of the more insightful reviews of the changing nature of positive externalities, arguing that chain migration can be understood as a diffusion process, which resembles an S-shaped curve (Faist 1997: 210).

The positive benefits of migrating to existing concentrations of migrants are low to begin with because the community is quite small. However, as the community increases in size, the benefits also increase and the trajectory of the curve rises steeply. Later, the advantages grow more modestly, the curve flattens and eventually, as net disadvantages set in, the curve starts to dip, gently at first but then rapidly. In other words, the potential social capital offered by ethnic enclaves, or migrant concentrations, to mediating risks increases slowly, then rises sharply as the migrant community grows, and then declines over time, as the nature of that community itself changes (Haug 2008).

INSURANCE AGAINST RISK

Probably the best known form of intermediation of risk in everyday life is the provision of commercial insurance services. Insurance companies provide insurance against a whole series of risks. For companies, these may include insurance against being sued by customers, or insurance against inclement weather. For individuals, it may include insurance against accidents, ill health, the breakdown of domestic appliances or a car, or the loss of personal possessions or a house. Travel insurance is a particularly well-developed market, and individual tourists can purchase insurance against a range of negative risks and uncertainties, including having to cancel a holiday due to family bereavement or personal illness, health insurance to pay for care abroad or for transport to return for care in the country of origin, and theft or accidents during the holiday. The insurance industry commodifies the provision of remedial action in the face of such risks, and this is achieved by their collectivisation across larger groups of individuals (insurance premiums) and by offloading some of the risks onto the reinsurance industry. Their charges are based on the estimated risks of particular events occurring, based on existing knowledge, and the profiling of individual purchasers of insurance. As Cunliffe (2006) contends, 'The modern science of risk calculation and determination is evident in the rationalist, actuarial tables of insurers and risk analyses by those in the business of risk' (Cunliffe 2006: 31). They cannot remove the uncertainties and risks associated with travel, but they can offer some certainties associated with negative events: replacement of stolen goods, payment of hospital bills, reimbursement of travel costs.

Migrants are likely to purchase a range of insurance policies at the destination, including the usual car and house/household insurances that are purchased by the indigenous populations. Sometimes they have to pay higher premiums or rates than the indigenous population, reflecting their lack of residence history (that is, documented history of relevant behaviour) in that country. They may also have to pay higher fees for car insurance if they have, or are perceived to have, higher risks of accidents (Box 7.3). There are also specific concerns focussed on the rights associated with different migrant statuses, most notably access to health services and unemployment

> **Box 7.3   Migrants as Risky Drivers and Car Insurance**
>
> In the UK, in 2010, the Admiral Group insurance company was found to charge migrants more than non-migrants for car insurance. The Admiralty Group stated that they charged migrants approximately 20% more for their policies because they received approximately 20% more claims from them. In other words, the rates reflected their calculation of risk probabilities.
>
> The story was taken up by the *Daily Telegraph* newspaper which reported Kevin Delaney, the road safety chief at the Institute of Advanced Motorists and a former head of traffic at Scotland Yard, saying that some newly arrived migrants lacked knowledge of UK traffic laws, and also had different attitudes to safety—and hence to risk:
>
>> A lot of people coming to the UK, especially from Eastern Europe and some of the African states, are coming from countries or backgrounds where there is a completely different attitude towards safety.
>
> This apparently minor story reflects many of the themes we engage with in this book. The insurance company adopts a modernist approach to risk, which considers this to be known and calculable (see Chapter One). The migrants are assumed to have either greater risk tolerance, or less competence to manage risks (see Chapter Three). And the *Daily Telegraph*, via its particular reporting of this event, contributed to the production of discourses about migrants being sources of risk (see Chapter Nine).
>
> *Source*: Authors' commentary based on a report in the *Daily Telegraph*, 22 August 2013. www.telegraph.co.uk/news/uknews/immigration/7458100/Migrant-drivers-charged-higher-insurance.html

benefits. Access to these varies across countries, especially in relation to the balance between state and private sector provision of such services. In some cases, countries may have bilateral mutual recognition or mutual provision agreements, whereby citizens of other countries are provided with similar or equal health care access to that available to their own citizens. Perhaps the best known, and most extensive, of such schemes is the EU health space, where, initially, individuals could acquire an E111 form which entitled them to health care services in other member states; this was later replaced by the European Health Card.

Another source of variation is migrant status, and whether an individual is classified as an irregular migrant, visitor, temporary resident or permanent resident/citizen. Many migrants may pass through a number of changes in their migration status across the migration cycle. Of equal concern are their rights to continued access to health care in their home country, which may depend on where they are registered as permanent residents. This is a particularly difficult issue for those who lead transnational lives.

The access rights of, and provisions made by, older migrants has attracted particular research attention, perhaps because of their increasing need for health care services as they pass from the active elderly to the frail elderly stages of the life cycle. For example, Böcker (2008: 105) studied the 'social security strategies' of later-life retirement migrants to Spain and Turkey from the Netherlands, as they try to balance acquiring or retaining access to a range of public and private resources in their countries of origin and the destination countries, especially pension and health care rights. These strategies take a variety of forms, depending on the resources available to them, and may include: retaining a house in the country of origin, registering as being resident at the home of a relative in the country of origin, or regular return trips in order to fulfil particular registration requirements.

King *et al.* (2000: 189–192) provide a detailed case study of the practices of retired British migrants living in southern Europe. They found that on average 59.5% of the retired British living in four Southern European regions possessed health insurance, with this being lowest in Tuscany (45%) and highest in the Algarve (69.7%). They had obtained insurance to cover a number of different forms of health care access, and in ranked order these were: local hospital care costs (48.7%), UK hospital costs (37.0%), local nursing home fees (27.3%), UK nursing home fees (19.4%) and costs of travel to the UK for treatment (15.2%). The fact that 40.5% had no insurance reflects two possibilities: first, and most likely, they were eligible for public sector health care in the destination, which depends partly on age and partly on having registered locally; and secondly, that they were exposed to the risk of having no access to a range of health care services, other than emergency care. This second category is probably quite small, and is either very risk tolerant, overconfident about their health prospects, or have sufficient resources to pay for health care at full market cost, if the need arises.

International retirement migrants constitute only a small proportion of all international migrants, and Ku (2006) provides a broader overview of the position in the USA, where the private sector dominates health care provision. This study found major differences in health care insurance between three categories of persons. Not even all US-born citizens (only 86.7%) and naturalised citizens (82.8%) had health insurance coverage in 2005, but the proportion was less than one half (43.9%) amongst noncitizen immigrants, which of course exposed them to greater risks. Recent and irregular immigrants were more at risk, having the lowest levels of health insurance coverage. Over time, as migrants obtain better paying jobs (reflecting the acquisition of nationally specific human capital), rates of health insurance increase.

There are two main reasons for the differences between the different categories. First, immigrants were less likely to be in receipt of employer-sponsored health insurance because of the types of industries they worked in: agriculture, construction and hospitality have low levels of provision.

Secondly, the gap persists even when only considering those in low-income jobs, because immigrants were less likely to be eligible for public health schemes such as Medicaid and Medicare, which serve the poor and the elderly. This dates back to the 1996 welfare reform law which excluded most regular, permanent residents (but noncitizens) from receiving Medicaid and other welfare benefits during their first five years in the USA. Migrants, therefore, are even more in need of private health care insurance, but the costs are prohibitive. There is also a vicious circle, because uninsured migrants are more likely to have poor health, and this makes them more likely to be unable to work and generate income to pay for health care (Ku and Matani 2001). Some migrants may also prefer to maximise their savings, while others simply lack sufficient resources to pay. The risks are severe because, lacking health insurance, the costs of even a single period in hospital can lead to indebtedness. In these circumstances, the migrants have to rely on 'safety-net clinics and hospitals', usually run by charities or public bodies, that provide free or low-cost health care.

Irregular migrants are most at risk in terms of access to health care. In Europe, the International Covenant on Economic, Social and Cultural Rights ensures emergency health care to undocumented migrants as a basic human right. In most European states, this is the only form of provision they are eligible for. However, some countries, such as France, the Netherlands and Spain, provide health care access to undocumented migrants which is equivalent to that provided for their own citizens (Kemefick 2013). For example, prior to September 2012, undocumented migrants in Spain who registered with their local municipality were eligible for full coverage through the public health sector. In practice, however, some lacked appropriate documents to register for these services. There was also a fear that registration could result in police or immigration authority measures against them. The position in Germany is particularly complex. Irregular migrants can access emergency care without risk of being reported to the police or border authorities. However, they can only access other services if they acquire a medical card, and the authorities which issue these are required to report irregular migrants.

In the face of the difficulties experienced by migrants in accessing commercial and public insurance schemes (and health care provision), there has been growing interest in the potential to utilise microinsurance (Power *et al.* 2011). Box 7.4 summarises examples of microinsurance schemes. Microinsurance schemes come in many formats, and this study reports on the schemes used by Mexican migrants in New York City. It identifies three models for microinsurance: home, host and hybrid models. In practice, most schemes are home country models, and, because they are restricted to a particular nationality, remain relatively small scale. Host country models are rarer, but are likely to be larger scale. Taken together, they have limited coverage, so that there was considerable unmet need amongst the Mexican

---

Box 7.4   Examples of Microinsurance Schemes

- **El Salvador.** Seguros Futuro is a cooperative insurance company which provides insurance products to migrants that will cover their repatriation costs (for example, if they become unemployed or ill and unable to work), as well as the maintenance of year-long remittances in case of death.
- **Indonesia.** The state requires migrant employment agencies to provide insurance packages that cover a range of risks: death, disability, medical expenses, unpaid wages, deportation and physical abuse. Some 95% of migrants are estimated to use these insurance packages.
- **Spain.** The SegurCaixa company, which is affiliated with the La Caixa cooperative, is a rare example of a host country insurance company designing and providing specific insurance policies for migrants. Some 80,000 migrants, mostly from outside the developed countries, purchased these insurance policies in 2008, which insured the following: death insurance, repatriation of the migrant's remains, and regular monthly income for the families of the deceased for a period of five years.

Source: after Power *et al.* (2011)

---

population for risk mediation products to cover risks both in the home and host countries. These include insurance against accidents, repatriation, protection of remittance flows, and cover for family living in the home country. Meanwhile, ironically, it seems that while migrants may constitute part of the risk-mediating strategies of households in their home countries, they themselves are often unable to secure adequate insurance cover for their own needs.

There are a number of reasons for the poor provision of insurance coverage amongst the New York Mexican community. Above all, there are legal and regulatory barriers because home country providers in Mexico are prohibited from selling their policies in the United States, while standard American insurance companies lack adequate low-cost distribution channels to reach that community, and there are virtually no host country microinsurance providers specifically for migrants. Power *et al.* (2011) argue that these obstacles can only be overcome through closer collaboration with private and public insurers in the host and home countries, as well as the Mexican community. There is also a need for new forms of distribution, through sports and other community organizations, and new forms of low-cost payment services, to allow microinsurance companies to thrive for the Mexicans in New York, and for other migrant communities. One option is to work in partnership with financial transfer agents, as these tend to be trusted by migrants.

EMPLOYMENT AND MIGRATION AGENCIES: MITIGATING THE RISKS OF JOB FINDING AND TRAVERSING BORDERS

One response to the existence of both systemic and non-systemic risks in migration has been the growth of various forms of intermediaries which seek to mediate, while commodifying, these risks. This applies to both intermediary-migrant and intermediary-employer relationships. Migration agencies are not a recent intervention in the migration process. They have long played a part in facilitating migration, by reducing the costs and risks associated with transborder mobilities. Initially, these were often government bodies, or specially commissioned agencies, but latterly they have become more commercialised. Baines (1995; 1998) provides an account of the historical evolution of assisted or intermediated migration relating to Europe. As early as 1846–1869, there were some 339,000 officially assisted emigrants from Europe, mostly destined for the expanding European colonial empires, or for the Americas. In later years, the role of mediating risk passed from European governments to destination governments, or quasi-governmental agencies. For example, a Canadian government scheme led to the migration of some 200,000 Ukrainians to Manitoba in the years1896–1914 (Marr and Paterson 1980). Baines (1995: 47) comments on the outcomes of these intermediations:

> If settlement schemes reduced the total emigration costs for individuals, we would expect that the rate of emigration would increase. But a more important effect was probably to divert the migrants from one destination to another.

If Baines is correct, this suggests that factors other than the costs and risks associated with migration were more important in who migrated. It is difficult to test these ideas in a historical context, and it is perhaps more useful to contemplate that migration agencies may influence the migration level, who migrates, and their destinations.

What exactly is the role of migration agencies? Most obviously they reduce the costs and risks associated with migration, in a number of ways. The starting point for understanding their role is the knowledge asymmetries that exist between migrants and intermediaries. Agencies have resource advantages, including a comparative knowledge advantage, over most migrants in relation to understanding and mediating risks. They benefit from economies of scale, being able to spread the costs of knowledge acquisition across their migrant clients. Previous involvement with earlier migrants has also meant they have acquired not only codified but also tacit knowledge of the migration process. This still leaves the issue for the migrants of how to decide which agency to use. The Four Corners agency for example (Box 7.5) emphasises its high rate of success. However, unless you have a personal recommendation from a trusted friend, then a different

Box 7.5   Four Corners Emigration Agency: Reasons for Using Migration Agencies

The Four Corners agency emphasises the advantages of using the advisory services of a migration agency to secure a visa:

> The decision to emigrate to a new country is possibly the biggest decision you'll ever make, so you want to be sure that you obtain the visa that will enable you to achieve this objective. You can reduce the possibility that your application may be returned, refused or rejected or delayed.

They also express this in terms of risk:

> A proportion of applicants to our destination countries consult a lawyer or advisor at some stage in the immigration process to achieve their emigration objective and to protect themselves from the risk of making a technical error in the visa or residency process.

Four Corners specialises in putting potential migrants in touch with recommended advisors. They state that these agents can provide 'the things that matter most to you and your family when you are emigrating':

- Referral to your own independent advisor who contracts directly with you.
- A high application success rate.
- Expertise in visa processing and resettlement.
- Personalised, in-depth assessment procedures.
- Ongoing support and assistance from start to finish.
- Peace of mind.
- Help to protect your interests in the emigration process.

In other words, they promise that they can decrease the risk of failure (they have high success rates) by drawing on their expertise (tacit and codified knowledge), thereby providing reduced uncertainty (peace of mind).

Source: www.fourcorners.net/whyuseanagent.php

source of trust is needed in the face of uncertainty about the reliability and effectiveness of individual agencies. Some form of official recognition or registration of agencies can provide a source of formalised trust. This is emphasised on the Australia Government Department of Immigration and Citizenships website, which offers the following advice: 'You do not need to use a migration agent to apply for any visa. However, if you choose to use a migration agent, you should use a registered migration agent. Hear from an unsuspecting visa applicant who used an unregistered agent, lost all her money and still ended up without her visa . . .' (www.immi.gov.au/visas/migration-agents/unregistered-agents.htm).

Migration agencies may either work directly with migrants, helping them to secure visas (as in Box 7.5), or act as intermediaries between migrants and potential employers, reducing the costs of job finding and recruitment, respectively. In effect, they also reduce risks for both parties by filling the information gap—sometimes a chasm—between the employers and migrants. The role of migration agencies is partly determined by national migration regulatory regimes. The more complex and exclusionary the regime, the more likely it is that there will be a role to be played by such agencies. For example, in the 1980s the UK government sought to push visa controls back from the national border to the pre-entry stage, that is, to the countries of origin (Flynn 2005: 467). Combined with restrictive migration controls, this meant that many potential migrants increasingly sought the assistance of migration agencies.

Employment agencies tend to specialise in either the high-skilled or the less-skilled ends of the labour market, as Dench *et al.* (2006) report in the UK. Headhunters are sometimes used in the highly skilled market segments as a way of prefiltering candidates, rather than openly advertising posts. They undertake a first sifting of likely candidates, but interviewing in the later stages is usually undertaken by employers. These agencies are not exclusively dedicated to recruiting migrants, but they are increasingly likely to work with migrants, contributing to the internationalisation of labour markets. They have a number of advantages in terms of their knowledge of potential applicants, networks of informants about applicants, and knowledge of foreign qualifications.

Recruitment agencies in the less-skilled market segment focus more on delivering units of labour, than seeking out the most highly skilled individuals to fill particular posts. They tend to be used by employers to fill specific gaps in the indigenous supply of labour, whether these are related to particular skills, industries with difficult working conditions, or seasonal work. In some cases, they pay particular attention to language skills and qualifications, but in others—such as agriculture—it is more a case of body counts: so many bodies to undertake so many days or weeks of work, in order to forestall the risk of labour shortages. Employers may sometimes develop close working relationships with particular agencies, again reflecting the importance of trust in mediating risks. Dench *et al.* (2006: 45) report that: 'One horticultural employer reported a good relationship with the local Jobcentre for filling seasonal jobs—the account manager visited regularly and understood the type of jobs offered and employees needed'.

The agencies do advertise their services, but word of mouth—meaning recommendation and trust—are also important, as Dench *et al.* (2006: 43) report:

> It soon becomes known among the migrant community which employers are recruiting and which offer the better terms and conditions. An employment agency in the *Administration, Business and Management*

sector commented: 'They come to us. We are lucky we do not have to advertise. People seem to know when it is our busy time and apply'.

Migration agencies also provide other forms of support for potential migrants, including help in securing visas, and opportunities to acquire tacit knowledge about living in the destination country: 'Once applicants have been offered a job, they attended a seminar about the UK and what it is like to live there' (Dench *et al.* 2006: 43). This may be facilitated by employing migrant workers to act as mentors for the new migrants—again, in order to utilise tacit knowledge, with the aim of minimising risks, and turning uncertainties into risks.

Migration agencies are especially likely to be used by first-time migrants. Their migration experiences will usually provide these individuals with learning opportunities, and the tacit knowledge acquired will reduce the risks and uncertainties associated with future migration. In the terminology of behavioural economics, they acquire enhanced competences to manage their migrations. Hence, sequential or repeat migrants are less likely to rely on migration agencies, making direct contact with employers instead. Dench *et al.* (2006) report on how agricultural employers in the UK who had first employed particular migrants via migration agencies were subsequently approached directly by these individuals when they wished to secure further job contracts. This was especially likely in the case of migrant workers from the EU accession states, who did not need assistance to acquire visas or work permits.

Williams and Baláž (2004) report a similar shift from reliance on agencies to self-organisation in the case of au pairs from Slovakia working in Western Europe. At the time they undertook their research, there were some 200 au pair agencies in Slovakia that specialised in international placements but, because many of these were dormant, a relatively small number of agencies managed most placements. They constituted part of the 'commodification of the transnational maid trade" (Momsen 1999: 9). In their study, the authors asked the au pairs to rank the importance of alternative recruitment channels. The au pair recruitment agencies were first ranked, followed at some distance by self-organisation. This is broadly similar to the findings of the Bratislava Centre for Family Studies (Hajduchova 2003), which found that 71% of au pairs used specialised agencies, and only 29% organised their own placements. The agencies can mediate the risks for both the migrants and the families who employ them. They also offer a form of safety net for the au pairs, as they provide mechanisms to assist them to change host families. Of course, they cannot guarantee against unsatisfactory placements, or even abuse, but they help minimise such risks. However, even while abroad, there was evidence that networking amongst other Slovaks in the same countries meant that the migrants became more self-reliant, and were able to arrange their own moves to different families; such networks had been used by approximately a quarter of those interviewed.

The Impact of Internet on the Use of Agencies: Disintermediation or Reintermediation

The growth of the internet has challenged the relationship between migrants and traditional intermediaries, such as recruitment and migration agencies, as well as traditional networks. This has two main expressions: internet reintermediation, and social media usage.

First, the internet allows migrants to acquire significantly more codified knowledge about destinations, jobs and travel, and also about migration regulations. They can read online about the migration requirements of particular countries, about jobs abroad, and about some of the risks of living and working abroad, at least as reported in a large number of Internet sites, including newspaper web pages. There is a transaction cost attached to this, in the time taken, but it has created the potential to find a substitute source of knowledge in place of that held by agencies. There are limits to this, however, because there are constraints on the extent to which tacit knowledge can be translated into codified knowledge (Nonaka and Takeuchi 1995) on websites. Nevertheless, an increasing number of migrants are directly managing their migration abroad, either through web-based intermediaries, or directly engaging with employers, consulates, etc., via the Internet. Moreover, as the au pair example (Williams and Baláž 2004) indicated, migrants are especially likely to manage subsequent migrations themselves, where their first migration experience has provided tacit knowledge of the destination, and of the implications of being a migrant.

Where they use the internet to find an (commodified) advice centre, such as the Four Corners website (see Box 7.5), this is not so much a case of disintermediation of the migration process as a reintermediation; they are substituting new types of agencies for the traditional ones with physical offices, and face-to-face contacts. Whether they are paying for this advice, or the commodification comes directly or indirectly in the advertising revenues of the search engines they use, it is still a case of reintermediation, and their reliance on the superior, but necessarily selective, knowledge and practices of the intermediaries.

Secondly, social media websites provide an opportunity for direct interchanges of tacit and tacit-to-codified knowledge (but also opinions, values and experiences) amongst migrants and potential migrants relating to migration risks and uncertainties. These can help to convert uncertainties into risks, or identify hitherto unimagined uncertainties, as well as suggesting ways to mediate risks. Social media sites are particularly important in migration because they provide opportunities for exchanging tacit knowledge via, say, Skype, or codified knowledge via Facebook. There are limits to the extent to which tacit knowledge can be captured by such media, but they provide unprecedented opportunities for widening the range of individuals (whether friends, acquaintances, or previously unknown individuals) that the potential migrant can learn from.

The increasing use of the Internet has important implications for migration and risk, whether the migrants rely on Internet search engines to locate codified knowledge and agencies, or social media to exchange tacit and codified knowledge. The increased access to knowledge may lead individuals to become more competent in managing risks, or at least to believe that they are more competent. If so, this may feed into a higher migration propensity, or a greater willingness to 'migrate to more risky destinations'. For example, indirect evidence on this is provided by Umblijs (2012), who examined the relationship between network size and the probability of migration: the larger the immigrant community is in Germany, and presumably the larger the associated migrant networks, then the more likely it is that more risk-averse immigrants had been attracted to Germany in the period 1960–2000.

SMUGGLING AND TRAFFICKING: CONTRASTING RISKS

In contrast to relying on legalised and regulated agencies, migrants may also utilise agencies which are operating in the grey areas of regulation, or completely outside the regulatory regime. These attract far more media and policy attention because of the risks they are seen to pose to the migrants themselves, and to the integrity of the immigration regimes of particular countries. They are epitomised by smugglers who offer, at a price, to reduce—or more realistically, transform the nature of—the risks associated with illegal border crossings (Koser 2008). In general, the role of smugglers increases in relation to the degree of border closure (Väyrynen 2003). It is no surprise then that, as governments tighten national migration regimes, so the scale of smuggling tends to increase. Koser (2008), for example, reports some sources as estimating that, in the mid-2000s, up to half a million people a year were being smuggled either out of or into Pakistan, with many being in transit.

The United Nations Office on Drugs and Crime (UNODC 2010) defines smuggling as: 'Smuggling migrants involves the procurement for financial or other material benefit of illegal entry of a person into a State of which that person is not a national or resident'. As with migration and employment agencies, smuggling is based on knowledge asymmetries. Migrants who are denied regular access to a particular country may lack the knowledge required to try to enter as irregular migrants. This can range from bribery of officials, to knowing paths across mountains, to having access to forged papers, to owning means of alternative transport. Smugglers have the knowledge and contacts to provide access to these irregular channels, or at least they claim to. In effect, they offer to reduce the risks the migrants would be exposed to if they tried self-organising irregular border crossings, but this does not mean that they eliminate such risks. Smuggling is a profitable and large-scale business for the smugglers. Migrants may pay as much as $70,000 to be smuggled from China to the USA (Petros 2005), and this single route might—by extrapolation—generate as much as $100 million per annum for the smugglers.

Other routes, involving shorter distances or less carefully guarded borders, or perhaps higher risks to the migrants, may be far less expensive. Vullnetari (2012: 130), for example, has described how Albanian migrants used smugglers to help them enter Greece and Italy. One of her interviewees told her how, in the late 1990s:

> 'We would walk for three to four hours through where the barbed wire used to be to cross into Greece, over the mountain . . . together with women and children . . . the smuggler would wait for us in one of the border villages. He was connected to the guy who had the car . . . and they sorted between themselves the money we paid them . . . we paid 80,000 drachmas [Euro 250] per person to include all these things and they would bring us to Thessaloniki.

Another, and more recent, migrant went from Turkey to Spain to Cuba, but failed to find a way to be smuggled from there into the USA. So he returned to Albania, and tried again, this time successfully. The total cost was about $30,000 (Vullnetari 2012: 130). Migrants are usually faced with a choice of options, with the more risky ones being lower costs. For example, Koser (2008: 11) found that smugglers of migrants from Pakistan to Europe used a range of routes: flying directly, flying via a transit country, by sea, or overland—or some combination of these. Although there are cost differences, most forms of smuggling are relatively expensive. Sometimes the migrants may have accumulated sufficient funds to pay the smugglers, but frequently they or their families will have had to borrow substantial sums, on the basis of this being an investment, to be repaid by the higher income the successfully smuggled migrant is expected to earn.

Koser (2008) provides an illuminating account of the smuggling of migrants from Afghanistan and Pakistan to the UK. Some were able to pay the smugglers from their family savings, others sold land or possessions, and others borrowed from money lenders. Because the risk of deception is high, the payment was usually made to a third party, 'and in effect negotiated a "money back guarantee" with smugglers. This system concurrently increased the risk for smugglers, who invested substantially in the process without a guaranteed return' (Koser 2008: 17). Most of the migrants who took these routes usually found a job relatively quickly and, on average, after two years their remittances had covered the fees paid to the smugglers. The migrants ended up with higher incomes, the incomes of their households were doubled, and the smugglers made a sizeable profit through the managing of risks. Other smuggled migrants had far less happy, and sometimes tragic, experiences, and found themselves exposed to high levels of risk by incompetent or deceitful smugglers.

Smuggling is usually constituted of a chain (Bilger *et al.* 2006; Neske 2006). Salt and Stein (1997: 467) capture this in terms of it being 'a system of institutionalized networks with complex profit and loss accounts, including a set of institutions, agents and individuals each of which stands to make a commercial gain'. Indeed, Koser (2008: 3) reports that the smugglers in his study redistributed about one half of the smuggling fee to other members in the network, such as forgers, suppliers of passports, and migration and transport officials. There are, of course, risks for the smugglers as well as the migrants, although of a different nature and order. But, as Koser (2008: 23) argues, the existence of a network mediates the risk for the smuggling process: 'As with all networks, risk is effectively spread through the network. Arresting any single operative is unlikely to be anything more than a temporary setback for the functioning of the network as a whole'.

The chain starts with the recruiters. The recruiters act as intermediaries between the migrants and those who will actually transport them across borders, although the chain may also involve other agencies such as money lenders, document forgers, and the owners or drivers/captains of boats or lorries. The key to recruitment is winning the trust of potential migrants, which means that recruiters tend to live in the country of origin or the transit country, and have at least a reasonable working knowledge of the migrants' language. Sometimes they know them personally, and come with recommendations from friends and family who have previously been smuggled by these agencies. On other occasions, the migrant—perhaps trapped in a transit country—has no option but to trust smugglers whom they know little about. In these circumstances, 'Recruiters prey on vulnerable persons and exploit their vulnerability. They will often tempt people into migrating, often misinforming them about both the process and the reality of the destination country' (UNODC 2010: 14). There are, therefore, a range of smugglers, and smuggling conditions, with variable degrees and types of risk.

At one extreme are the tragic deaths of migrants smuggled by small boats or in overcrowded lorries (see Box 7.6). But the risks of crossing borders are manifold, as Collyer (2010) reports in the case of African migrants travelling to the Mediterranean coastline, as a staging post for journeying to Europe. They have to cross seas, deserts and mountain ranges, either through self-organisation or relying on smugglers, but in both cases the risks are substantial. The risks, however, are magnified whether from being robbed, being cheated by smugglers or being abused and enslaved by traffickers. There were also organised gangs who extorted money from the migrants. They are highly vulnerable, because of the difficulties of turning to the authorities for assistance.

The risks that migrants are exposed to in the course of smuggling can be considerable. At one extreme, it may involve purchase of false documents and travelling on regular transport links, or a short crossing at night in a relatively unguarded border zone. These carry risks of course: of detention for a short or long period, of loss of the investment they have made in the smuggling process, of being returned to their country of origin or the last country they left. But the risks can be far greater than this, for smuggled

Box 7.6   Risking Death: Irregular Migration From Africa to Spain's Canary Islands

On 15 May 2007, the Inter Press Service Agency reported that: 'A new wave of African immigrants has left a number of victims behind on the route to Spain's Canary Islands, where 1,300 undocumented immigrants have arrived in the last five days'. Amongst these were 28 migrants whose bodies were found washed up on the coasts of Western Sahara.

Unnamed Spanish government sources commented on the role of smugglers and traffickers in these migrations: 'It's no longer a question of a make-shift craft trying to reach our shores as best it can, but a complex operation involving synchronised movements of vessels'. There was an example of this on 11 May 2007, when 11 boats arrived within a few hours on the southern coast of Gran Canaria. The migrants had been at sea for some 25–30 hours, crossing 225 kilometres of open sea. Near the end of their journey the boats dispersed so as to reduce the risk of being detected. Typically they paid between $1,000 and $1,500 dollars for the trip.

One of the migrants explained the obvious economic attraction: 'I came because here you can get a much better job for much more pay than the work we did as children in my country, either because our parents sent us out to work, or to earn ourselves a few cents'. And of course he knew there were risks: 'Yes. Everyone knows. But what is worse, to take that risk, or to live a life that is very similar to death?' In other words, they think they know the risks. Another newly arrived described his fear during the rough sea crossing, saying that it was 'utter madness'.

Aminata Traoré, previously a government minister in Mali, placed this dangerous irregular migration in context of global regulatory systems:

> White people and their products can circulate freely, and they are the only ones who can come to Africa with their trade agreements which impose sanctions on African countries that refuse to let foreign companies in . . . while the doors to immigration are closed, and people are even being selected, like raw materials are selected. This is the great paradox. Europe takes everything it wants from Africa, but then it creates a barricade to keep out people from Africa.

Source: Inter Press Service News Agency website. Accessed 30.08.2013. http://ipsnews.net/news.asp?idnews=37744

migrants are highly vulnerable and exposed to risks, as explained by the UNODC (2010: 2);

> Smuggled migrants are vulnerable to exploitation and their lives are often put at risk: thousands of smuggled migrants have suffocated in containers, perished in deserts or drowned at sea. Smugglers of migrants often conduct their activities with little or no regard for the lives of the

people whose hardship has created a demand for smuggling services. Survivors have told harrowing tales of their ordeal: people crammed into windowless storage spaces, forced to sit still in urine, seawater, faeces or vomit, deprived of food and water, while others around them die and their bodies are discarded at sea or on the roadside.

Traffickers are also involved in moving people across frontiers, but this is enforced movement. The UNODC (2010) describes trafficking in the following terms: 'Human trafficking is the acquisition of people by improper means such as force, fraud or deception, with the aim of exploiting them'. There are times and places where smuggling and trafficking blur into each other, and the UNODC (2010: 2) describes the distinctions as being subtle and overlapping. Migrants may pay to be smuggled, but then during the process of crossing the border they may find themselves becoming subject to trafficking. This illustrates the double risks that individuals are subject to when being smuggled.

There is no doubting the risks that the migrants are exposed to, and their severity. However, there is an alternative view that smuggling represents—and is seen by the migrants themselves—as a rational economic decision, to invest in risk reduction in order to secure a higher return for their labour. As the work of Agustin (2007), Mai (2009), and Anderson (2008), amongst others, indicates, the experiences of migrants are diverse, and the degree of self-determination and agency is variable even in such precarious experiences as migrant sex workers. And the motives are not all economic. People are also smuggled in order to join families, or to reach a safer place (Herman 2006); but they all use smugglers to mediate risk—hopefully to reduce it, sometimes simply to transform it into other types of negative risks.

CONCLUSIONS

Migration is a relational process and it is defined and shaped by migrant networks, which can be a means for transmitting knowledge and assistance to potential migrants. As such networks are a means for reducing the risks attached to migration, and more importantly to translate some of the uncertainties surrounding migration into risks. This mediates who stays and who migrates, where they migrate to, and the channels of migration they use. Such networks are dynamic and their capacity to mediate risk also changes over time, as does their impact on the full transaction costs of migration. There is evidence of the differential importance of particular members of migrants' networks, with there being significant differences between family and kin versus different types of friends and relations. However, we still know little of the dynamics of migration networks and of how across the migration cycle, say, best friends become casual friends—with changes in their relational importance in terms of risk mediation—or the importance of different kin as sources of knowledge.

Migrants cannot always rely on informal social networks, however, so they may also rely on various intermediaries, many of which are commodified forms of providing assistance. These range from smugglers, to employment agencies, to visa application agencies. These can play a significant role, especially for first-time migrants, in reducing risks and uncertainties and to reducing the full transaction costs of the policy. But migration is also a learning process, so that over time individual migrants tend to become more competent at directly managing the risks of migration. These agencies range from the formal to the informal, and from government provided, to government regulated, to government (surveillance) avoiding. Migrants—especially regular migrants—can also employ commercialised insurance services which collectivise and redistribute risks, ranging from the health to the employment arenas. However, some forms of insurance cover are not available to migrants, and especially to irregular migrants, leaving them highly exposed to particular risks. A key issue for all migrants is knowing which intermediaries they can trust, given the costs involved and the potential returns on a successful migration, whether in terms of crossing borders, or of findings jobs or income.

Taken together these various forms of intermediaries effectively represent the meso level in the migration process, and often constitute the essential 'glue' that allows the migration process to function in the face of significant risks and uncertainties.

8   Migration, Society, Technology and the 'Risk Society'

INTRODUCTION

A number of writers, including Giddens (1999) and Beck (1992) consider that there have been qualitative changes in the nature of risk. There are new types of risk, and the sources of these have changed: they now include manufactured risks, generated by society itself, as opposed to external or natural risks (Giddens 1999). At the heart of such arguments is the changing nature of knowledge and technology, and the (in)ability to manage these and their unintended consequences. This is evident in a number of areas but especially those relating to health and the environment, with the latter addressing issues about environmental refugee movements in response to climate change. There have also been attempts to retheorise the relationships between societal and technological changes, and risk. This is not so much an argument about there being an increase in the scale and intensity of risk, but about the way in which societies are increasingly orientated, or even fixated, on risk. As Giddens (1999: 3) writes, 'It is a society increasingly preoccupied with the future (and also with safety), which generates the notion of risk'. Modernisation itself has generated hazards and insecurities (Beck 1992: 21) that have required systematic societal responses. The next section turns to be best known of all these theories of how societies engage with the changing nature of risk, that is, Beck's (1992) risk society thesis.

BETWEEN REALIST AND CONSTRUCTIVIST APPROACHES: THE RISK SOCIETY

Probably the best known of all the theories dealing with these societal shifts in the generation and understanding of risk is Beck's risk society thesis, which considers that risk is both real and 'out there' and socially constructed. It seeks to situate what he perceives as the increased importance, and changing nature, of risk in the context of changes in the very nature of society, and challenges to the dominant form of modernity. For Beck (1992), modernity was based on the scientific spirit of the Enlightenment and the internationalisation of economic activity which was nevertheless regulated by national states (Taylor-Gooby and Zinn 2005: 5–6). There were, of course, various forms of risk in modernism, and indeed prior to modernism (for example, the spread of plague by migration and trade), but in the late twentieth century, according to Beck, a 'risk society' emerged that was characterised by manufactured risks as well as by the traditional external or natural risks. This book contends that migration plays a significant role in generating and mediating such manufactured risks.

A key feature of the risk society is that the distribution of risks has been modified. Those who produce these risks (the wealthy, capital-accumulating class) are no longer immune from the risks that are produced from the application of knowledge to production. The owners of the capital which produces pollution, or pollutes food production, cannot isolate themselves from these consequences. These problems do not respect either national or social boundaries because, as Beck (1992: 36) expresses this, 'smog is democratic'. This is, of course, to some extent over-simplistic, as the wealthy have the resources to migrate more easily to non-polluted areas, or to buy imported foods, or to take other remedial measures. But in the risk society, not even the wealthy may actually be aware of the risks that are being manufactured. A sense of security is being replaced by a sense of uncertainty and vulnerability (Giddens 1990; Boholm 2003b).

Fundamentally, then, risk is seen as being the outcome not of too little knowledge, but of the application of knowledge that we can no longer fully understand and control. Therefore, the distribution of risk is not shaped by wealth but by the application of knowledge (Beck 1992; Ericson and Haggerty 1997). Of course, there were manufactured risks in the period before the risk society, but risk has now become an 'expression of highly developed productive forces. That means that the sources of danger are no longer ignorance but knowledge; not a deficient but a perfected mastery over nature' (Beck 1992: 183). According to Adams (1995: 181), the risk society is characterised by new technologies which flirt with catastrophe, unknown futures (cf. Knight's [1921] notion of uncertainty), and apocalyptic threats. Consequently, the risk society is focussed on how to prevent negative outcomes such as pollution and climate change rather than on producing positives such as increasing wealth and welfare (Taylor-Gooby and Zinn 2005: 5–6). Risks are being globalised and the uncertain outcomes 'cannot be handled through traditional methods of risk management', such as reliance on family support or insurance (Taylor-Gooby and Zinn 2005: 19).

An important component of Beck's work, and one which has particular relevance for understanding migration and risk, is the individualization thesis, which is very much a central tenet of his argument about reflexive modernism. Individuals are increasingly aware not only of the changing nature of risks, but also of their own potential roles as actors. Moreover, the decline in the levels of trust in industry, state and experts underlines the need for individuals to become more proactive in mediating risks, and shaping their own lives in response to these. As Taylor-Gooby and Zinn (2005: 5–6) comment:

> The breakdown of an established traditional order in the life-course provided by work, marriage, family and community leads to greater individualisation and increased uncertainty and anxiety. In this context, the individualised citizens of the world risk society are increasingly conscious of the responsibility to manage the risks they perceive in the context of their own lives, and, in this sense, self-create their own biographies.

Individualization is not just about the consequences of the decline of the traditional social order, an essentially negative effect, but also—and more positively—about individuals taking more responsibility for their life trajectories. It is an outcome of the emergence of greater reflexivity (Giddens 1990), in the context of a 'runaway world' (Giddens 1999), and the emphasis on 'active trust' as opposed to traditional deference, whether to authority embedded in class structures or to scientific expertise. Individuals, in effect, have both more need and more scope to create their own biographies, as part of Giddens's (1991) reflexive 'biographical project'. While this has a positive side, there is also greater uncertainty and fragility (Sharland 2006), for individualization is 'full of risks which need to be confronted and fought alone' (Bauman 2001: xvii).

Before we consider how the risk society conceptualisation, including the individualization thesis, can be applied to understanding migration and risk, it is first necessary to note some of the formidable criticism it has attracted. Some commentators fundamentally refute what they see as the doom-laden nature of the risk society, insisting instead on the continuing validity of notions of modernism and the enlightenment project as a source of science, development and progress (Adams 2001). For example, Wildavsky (1991) poses the question: 'If claims of harm from technology are false, mostly false, or unproven, what does that tell us about science?' Moreover, Beck's is not the only work to have considered the relationships between technology, culture and risk. Hughes (2004), for example, demonstrates that attitudes to technology have changed cyclically over recent decades, as reflected in shifts from the desire for technological fixes to economic growth, and concerns about the danger that technology 'can bite back' (Tenner 1997), if nature is treated in a cavalier fashion.

Above all, Beck's work has been critiqued for not being empirically grounded (Alexander 1996; Dingwall 1999). The evidence, in fact, is best described as mixed (Sharland 2006). Environmental and medical technology risks have become more globalised and less controllable (Matten 2004; Turner 2001). However, Hughes (2004) considers there is little evidence that technology is out of control, certainly in any specific and measurable way. Of course, there are examples that technology has been poorly tested and applied, and has hit back with substantial consequences for individuals, communities and states. Examples include global warming (discussed later in this chapter) and thalidomide. However, Hughes thinks that these are relatively insignificant (although clearly not for the individuals concerned) compared to the consequences of world wars, and the collapse of national states. Even the event at the Three Mile Island Reactor in the USA in 1979 probably did more harm to the reputation of nuclear power generation than to local communities. The evidence of a cultural shift to reflexive modernity has also been critiqued, for risk emerged as a major social category and focus for state intervention from the nineteenth century (Ewald 1986; see also Chapter Ten) rather than in the late twentieth century. Rose (1999) also contends that the emergence of the confident individual, creating his or her own biography, is far from universal and may only be found within particular social strata—social class, and the other traditional social orders, are far from being dead or irrelevant. For many critics, the risk society thesis underplays the evident persistence of deep social cleavages, in terms of gender, class and ethnicity (Furlong and Cartmel 1997). This is particularly important in this book because they are interwoven with migration as key structural determinants. The failure to recognise the importance of such structures is an 'epistemological fallacy' according to Furlong and Cartmel (1997), associated with the culture of individualism, and the outcome of neoliberal discourses (see Chapter Nine) that emphasise individual freedom to self-regulate. Migration is one articulation of such individualism, although it operates in an environment where neoliberal freedoms to mobility are in conflict with counter discourses about migration as a source of various risks (see Chapter Nine).

Perhaps the most significant problem of the risk society is that even if its broad brush approach applies reasonably well to catastrophic ecological disasters, it is far more problematic to apply it to everyday life (Tulloch and Lupton 2003), whether that be to risk taking by young people (Sharland 2006), or to migration. In short, although Tulloch and Lupton (2003) found there is greater perception and awareness of risk, and some evidence for individualization, 'the risk society thesis ... is not sufficiently situated, not sufficiently concerned with localised "tales from the field" ' (Tulloch and Lupton 2003: 128) in arenas such as migration.

MIGRATION IN THE RISK SOCIETY

Does risk theory provide a useful framework for understanding migration and risk studies? Tulloch and Lupton (2003: 128) considered this question in relation to localised 'tales from the field', and specifically referred to 'tales' about migration. They argue that because Beck's risk society is too focused on what we can term the mega scale, that is, on the 'cataclysmic democracy' of catastrophic environmental hazards, it cannot engage effectively with how societies deal with 'the risk associated with mass immigration' (Tulloch and Lupton 2003: 41). However, this is probably too negative an assessment. While the risk society thesis may not be able to explain the complex and nuanced discourses and policy debates surrounding migration (see Chapters Nine and Ten), it can provide insights into some of the framing conditions that are important for understanding some forms of migration.

Two of the most obvious examples relating to the implications for migration of the lack of control of new forms of knowledge and technology are climate change and health, which are discussed later in this chapter. Migration constitutes a potentially significant response to technologically driven climate change, as exemplified by the risks associated with potentially massive environmental refugee movements (Piguet 2008): risks to the individuals involved, and to the societies of origin and destination. Similarly, migration can contribute to the global spread of diseases facilitated by the compression of space and time, resulting in part from changes in (air) transport technologies. Other examples include forced migration in response to site-specific technological disasters, such as the Chernobyl nuclear power station meltdown. Although an apparently localised event, it was subject to 'a tendency to globalization' (Beck 1992: 13), with nuclear fallout materials being recorded thousands of miles away. This is a classic example of Beck's notion that wealth and power can no longer isolate particular groups from the disastrous consequences of environmental disasters: 'poverty is hierarchic, smog is democratic' (Beck 1992: 36). However, this is over-simplistic because how social groups respond to such risks still depends on their resources—in terms of wealth, access to transport, personal mobility, social networks, etc. There are major differentials in who is able and who is not able to outmigrate, and when they can move.

In other words, Beck's risk theory is unable to engage with the complexity of risk in everyday life (Tulloch and Lupton 2003). Migration and refugee movements have multiple determinants. These include technological changes and economic disasters, but they also include war, disease, famine and natural disaster, as well as a range of sociocultural considerations. Moreover, there is no evidence that migration has increased in the risk society. For example, the relative importance of migration—expressed as a proportion of the total world population—was broadly similar in the early twentieth century and the early twenty-first century (Chiswick and Hatton 2003). Technology changes do influence migration, but there is no simple correlation with broad migration trends.

Developments in transport following the application of steam power (mainly the railways and shipping) did contribute to condensing space and time in the nineteenth century, while the automobile played a key role in redefining individual mobilities from the early twentieth century (Williams 2013). These would contribute to climate change, but awareness of this and the consequent environmental refugee movements lay in the future. Air transport existed throughout most of the twentieth century, but its main impacts awaited the technological developments of jet engines, contributing to the era of mass international tourism from the 1950s. This also contributed to global warming, and facilitated the more rapid movement of people and contagious diseases across borders. However, the most significant impact of transport technology shifts, and especially organisational innovations related to low-cost carriers, came from the 1990s, revolutionising both the geographies of accessibility and the costs of air travel, with major consequences for temporary migration and other forms of short-term mobility (Williams and Baláž 2009). Increasing human mobility has been associated with drug smuggling, trafficking, and terrorism (Edwards 2009), but association is not causality. Are these the outcomes of a lack of control over technology and knowledge, or are they the outcomes of the inabilities of states to manage their borders in face of globalisation? These issues are difficult to disentangle, but we return to them later in this chapter.

This is not to say there is no relationship between migration and technological change, but that it does not fit within the risk thesis framework. For example, Collyer (2010) has discussed why migration to developed countries has become fragmented: long and dangerous 'fragmented journeys' have become increasingly common, involving long-distance overland trips across Asia or Africa in particular. These often end in migrants becoming 'stranded' in what were meant to be transit countries, such as Morocco. Of course, this is partly a response to the tightening of immigration controls, especially since the late 1980s (ECRE 2007), but technological changes are also important:

> Such lengthy overland migrations depend on the availability of the necessary communications and other technologies that facilitate movement, such as instant international money transfers along the route or the availability of cheap mobile communications. The prominence of trans-Saharan migrations since 2000, when significant restrictions on migration to Europe had been in place for more than a decade, but mobile phone access, money transfer facilities and widely available email were just beginning to roll out across the region, suggests that technological change is perhaps more significant in explaining the recent development of fragmented migrations than policy restrictions.
>
> (Collyer 2010: 4)

Moreover, Collyer considers there is an element of 'democratization' in the unforeseen outcomes of these technological changes. He argues that if increased controls were the drivers of increased trans-Saharan (rather than direct) migration, these flows would be dominated by those with more resources. And if technological changes were the drivers, this would facilitate greater participation by those with fewer resources. The evidence is that both the relatively wealthy and relatively poor now engage in long-distance overland migrations, although the very poorest are still excluded (minimum resource levels are essential). Consequently: 'Technological developments in the Saharan and Sahel regions have stretched sub-Saharan migration networks so that they now reach to North Africa and for some even into Europe' (Collyer 2010: 5).

Collyer's study deals with the unintended consequences of technological developments, and there are severe risks for the individuals who engage in such long-distance overland migrations. However, it does not fit into Beck's (1992) vision of cataclysmic risk as an outcome of too much knowledge. The next two sections consider two phenomena which can perhaps be seen in such a framework: climate change-related migration, and the relationship between migration and health risks.

CLIMATE CHANGE, RISK AND MIGRATION

Migration can be 'the result of short term ecological changes such as environmental catastrophes, but also of long term eco-demographic changes such as population growth or long term resource degradation' (Malmberg 1997: 35). These environmental catastrophes can be either natural or human made. Tsunamis and earthquakes are examples of natural disasters, while nuclear disasters or wars are human in origin. But this is too simplistic: although an earthquake or tsunami may be a natural phenomenon, its impacts are significantly shaped by human activity, including technological changes. For example, new technologies may have allowed large areas to be reclaimed from the sea, putting more people at risk from tsunamis.

Climate change is the human-influenced environmental change that has attracted most academic, policy and popular attention in recent decades. The exact causes, extent and speed of climate change are still contested, although there is increasing scientific consensus that human activity has made a significant contribution. At the heart of this is how the lack of knowledge, and control, of technological changes has contributed to global warming. Much of the impact of human activity on climate dates back to earlier technological shifts, especially the emergence of a carbon-intensive economy from the nineteenth century. It is not necessarily something that can be situated solely in respect of a more recent loss of control over technological changes in the risk society. But whatever the exact causes, migration is one of the most important responses to climate change (Renaud *et al.* 2007; Stern 2007).

Black *et al.* (2011a; 2011b; 2012) have reviewed the evidence for the incidence of both natural disasters and related migration flows. Although the evidence is problematic, a picture emerges of increasing risk of storms and floods, if not droughts, over the last three decades. In contrast, the evidence for climate change-induced migration is equivocal. One of the problems is that both the size of the populations, and the volume of insured property and economic assets in hazard-prone zones, have increased in this time period, and it is necessary—but difficult—to control for these factors.

Another difficulty is that data are more readily available for the much smaller-scale phenomenon of international migration than for intranational migration. While an estimated 3% of the global population are living in a country other than their country of birth, some 11% have been internal migrants (Black *et al.* 2011a). The authors conclude that, even without significant climate change effects, these figures would be expected to increase to an estimated 66 million additional international migrants, and 242 million internal migrants by 2050.

What evidence is there that environmental refugees—or migrants moving in response to environmental changes—have added or will add significantly to the already high levels of international and national migration? There are a number of estimates in circulation, some of which have gained considerable popular or policy credence (Black *et al.* 2011a). For example, Myers's (2002) estimate that there were 25 million environmental refugees in 1995, and that this number could increase to some 200 million by 2050, is quoted in the Stern Review on the Economics of Climate Change (Stern 2007). This is contradicted by other studies, for example Tacoli (2011), who considers that the migration implications of climate change will be relatively minor. These conflicting interpretations and estimates reflect variable methodologies, definitions of what constitutes migration, and poor data availability (Piguet 2010).

More recent research has sought to provide stronger theoretical frameworks for such analyses (McLeman and Smit 2006), and tends to emphasise that migration is only one of a number of strategies available to those living in areas impacted by environmental disasters: some stay, some migrate—and their migrations may vary in terms of length and duration. For example, Perch-Nielsen *et al.* (2008) conclude that floods will have relatively limited impact on migration, because of the availability of a range of adaptation responses, but that the consequences of sea level changes will be more significant because of permanent land losses. Black *et al.*'s (2012) own work situates the analysis of environmentally driven migration in context of the multiple drivers of migration:

> The factors that drive migration affect both the scale of migration but also whether populations decide not to move location at all. Most of the world's population are not and do not want to be migrants. . . . Even in the face of exposure to extreme environmental events such as in the aftermath of the Japanese tsunami of 2011, the great proportion of the population usually prefers to stay and rebuild.
>
> (Black *et al.* 2011a: S6)

Migration is only one response to climate change-induced migration, and where resources are available—which usually means the more developed countries—there is a propensity to adapt rather than migrate (Neumayer and Barthel 2011). Insurance companies also mediate some of these risks in more developed countries. For example, they may insure property owners

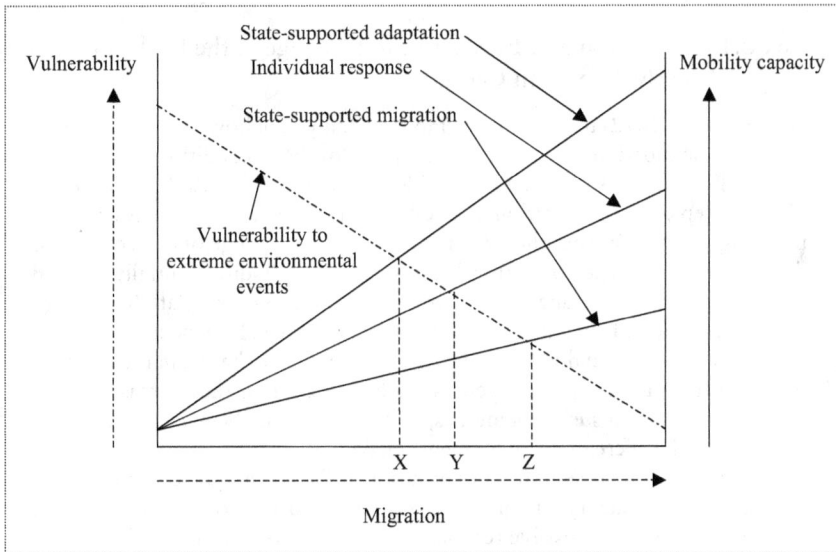

Figure 8.1 Vulnerability to natural disasters, mobility capacity and migration

Source: Authors

living in areas increasingly prone to flooding as a result of climate change, that is, by collectivising the risks. However, the economics of risk calculus may make market-led insurance policies prohibitively expensive, as the risk probabilities of flooding increase over time. State intervention may therefore be necessary to provide alternative state-subsidised insurance cover, or regulation to compel insurance companies to insure all properties at 'affordable rates'; the latter effectively means they will be cross subsidised by other insurance policy holders. However, less developed countries usually have poorly developed insurance markets, as well as limited state investments in flood relief schemes and other measures to mediate environmental risks. Figure 8.1, which draws inspiration from Black *et al*, (2012: Figure 3), summarises some of these relationships. The left-hand vertical axis represents increasing vulnerability, while the right-hand vertical axis represents mobility capacity, and actual migration is measured on the horizontal axis. There are three propensity-to-migrate curves, representing the existence of state support to facilitate migration, absence of any form of state intervention, and state support for adaptation. These produce three outcomes in terms of migration versus staying, with the proportion of migrants increasing from X to Y to Z for the three propensity-to-migrate curves.

It is difficult to establish causal relationships in this context, but Boustan *et al.* (2012) investigated the historical evidence of whether individuals respond to natural disasters by migrating or by adapting to changing environmental conditions in the USA in the early twentieth century (see Box 8.1).

Box 8.1 Migration and Environmental Change in the USA in the Early Twentieth Century

Boustan *et al.* (2012) undertook an empirical analysis of how the population in the USA responded to the 'natural disaster' of dust bowl conditions in the early twentieth century, as described so vividly in John Steinbeck's *The Grapes of Wrath,* an epic tale of migration from Oklahoma to California. The researchers used migration data from the 1920s and 1930s, and a range of environmental data on different types of 'natural' disasters. Their findings initially seemed somewhat puzzling: young male migrants did tend to outmigrate from areas that were prone to being affected by tornados, but they also tended to move to areas that experienced flooding. The explanation may either lie in the fact that flood control measures were seen as mediating risks in these areas, or that the public works on such schemes, especially during the New Deal years (late 1930s), may have created employment opportunities.

This research focussed on a period prior to Beck's risk society, but it does indicate the complexity of migration responses to environmental disasters. Migration is only one possible response, and adaptation is also possible, but in any case the outcomes are highly contingent on individual and state responses. The extent to which states have sufficient resources to support adaptation processes is also questionable. Black *et al.* (2012) consider that poorer countries are likely to face an 'adaptation deficit', and that migration is a more likely response to environmental change in these circumstances. Although such risks are hardly universally democratised, there may be some evidence that they are to some extent democratised within macro world regions.

Source: Boustan *et al.* (2012)

Rapid and substantial migration after extreme environmental events tends mostly to be intranational rather than international, although the latter is more likely to be important in small states, especially island states. McLeman (2011), for example, discussed the total abandonment of the small town of Plymouth in Montserrat following a volcanic eruption. The displacements are usually short-term, and individuals return after a relatively short time because there are risks in being displaced (e.g. moving away from land, from livelihoods based on tacit knowledge, and from support networks), as there are in remaining, or returning to, the affected area. Although they did not specifically address Beck's risk society thesis, Black *et al.* (2011a) also found little evidence of the 'democratisation' of risk in a risk society. Instead, they stress that environmentally driven migration has structural determinants, especially if both outmigration and return are considered. For example, the return of those displaced by Hurricane Katrina in New Orleans was strongly influenced by age and income.

The importance of contingent factors, which to some extent can be considered in terms of the full range of factors that influence migration, make it difficult to predict future migration responses to climate change. Black *et al.* (2011a) considered the available evidence against a multidimensional conceptual framework of the determinants of migration. They concluded that the main future risks in coastal areas are from the permanent loss of land and increased salinity due to sea level rises, and an increased vulnerability to flooding due to both sea level changes and extreme weather events, such as storms and cyclones. Changes in rainfall may also increase the risk from river flooding or drought, both of which can affect agricultural productivity, water supplies and the viability of settlements. Climate change can also cause land degradation, which effects ecosystem productivity, including agricultural productivity, runoff and flood risks. The effects are compounded by human activity, which has led to the loss of approximately one third of global mangroves and a fifth of global coral reefs (Agardi *et al.* 2005). This underlines the importance of both contingent factors and the other determinants of migration. For example, political drivers are also important, not only directly in terms of how oppression and conflict generates refugee flows, but also in terms of the capacities of states to respond to environmental pressures. McGregor *et al.* (2011) comment that droughts in Zimbabwe in the previous decade have had more severe effects on rural populations than earlier, more severe droughts because of the vulnerability stemming from economic collapse and political conflict. Environmental change impacts may be mediated by political drivers, but they may also be a cause of political conflicts exacerbating tensions over access to scarce resources.

Black *et al.* (2011a: S9) conclude that the effects of environmental changes on future migration will depend not only on changes related to climate change, but also on changes in the other migration drivers, which are 'inherently less predictable than environmental changes'. Of course, this is a macro-scale analysis, and whether individuals migrate is influenced by both personal and family characteristics, and also by the existence of barriers and facilitators. The personal characteristics are sociodemographic and socioeconomic, and the barriers are both regulatory and economic (See also Box 8.2).

Box 8.2   Migration and Climate Change: Adaptation Versus Displacement in Coastal Zones

Major threats from climate change relate to sea level rises, drought and desertification, and increases in the frequency of extreme weather events, such as storms, hurricanes, and floods. The UNESCO report on Migration and Climate Change (Piguet *et al.* 2011) estimates that about 602 million people live in 'low elevation coastal zones' (at an altitude of less than 10 metres above sea level). Droughts and desertification may 'only' affect about 2.2% of the land surface, but this is home to some 10.5% of the world population.

Several large developing and middle-income countries (India, China, Bangladesh and Vietnam) have densely populated areas, and some of their

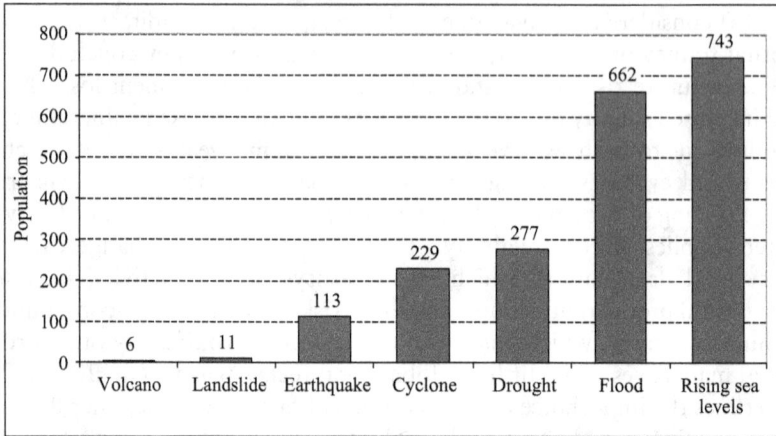

Figure 8.2   Population (millions) in urban agglomeration by type of natural risk: 8th–10th decile of risk

Source: UN World Urbanization Prospects: The 2011 Revision

highest quality agricultural land is in low-lying coastal areas. The UN World Urbanization Prospects database (United Nations 2012) contains a list of 633 urban agglomerations with 750,000+ inhabitants which are considered to be vulnerable to various types of natural risks (Figure 8.2). For example, there were 243 coastal agglomerations with a combined population of 734 million in the 8th–10th decile of (higher) risk from rising sea levels. Some agglomerations faced multiple hazards. There were some 88 agglomerations (with 299 million inhabitants) which were significantly at risk from two hazards, and 10 agglomerations (29 million) from 3+ hazards.

Rising sea levels, droughts and desertification alone may affect over 1.4 billion people. This poses the question of whether these threats will result in absolute population movements at an unprecedented scale in world history. Some alarmist predictions (Stern 2007: vi) estimate the potential number of individuals displaced by climate change to be some 200 million by 2050. This number is close to the current total of 214 million international migrants (United Nations 2012).

Forecasting the numbers of international migrants is notoriously problematic, and it is also difficult to quantify the effects of particular threats on migration flows. Some countries and regions may devote more resources to adapting to climate change. Increases in sea level, for example, can be countered, at a cost, by the construction of seawalls and barriers. Building long-distance ducts, constructing dams, introducing irrigation systems and changes in crop structures may help mitigate some impacts of droughts and desertification. Improved systems of weather forecasting and flood management may also alleviate the consequences of extreme weather.

Adaptation to new climate condition may offset some potential migration flows. However, some regions and countries may have very substantial, and perhaps insurmountable, difficulties with adaptation, with the consequence of large-scale population displacements. Not all displaced individuals become international migrants. Parts of the displacement may take the form of internal migration, especially in large developing and middle-income countries such as India or China. There are already strong internal migration flows in these countries, primarily driven by urbanisation and industrialisation processes. Climate change may contribute to future rural-urban drift, but it is difficult to estimate the extent of this.

Climate change, nevertheless, may increase the risk of competition for scarce land, food and energy resources, whether within or between countries. Where will the migrants go? Some authors (Newland 2011) assume that most migration flows will be informed by geographical proximity, historical (often colonial) links, and ethnic and linguistic ties. These ties and links already shape migration flows in Europe and the Americas. By the 2010s, migrants from Central America and the Caribbean headed for North America. Migrants from North Africa were crossing the Mediterranean, often at great personal risk (see Chapter Seven), to arrive in the southern EU Member States. It is also estimated that over 20 million illegal labour migrants and climate refugees from Bangladesh have tried to cross the Indian borders. India has already built a 3,406 km long, 3.6 metre high, double wire fence on the Bangladeshi border to control the flow of future forced climate migrants (IOM 2008: 40). Irrespective of the actual risks, which still remain somewhat unclear, the perceptions of long-term risk are already influencing individual and state responses to climate change.

Sources: Newland (2011); Piguet *et al.* (2011); Stern (2007); United Nations (2012)

Finally, drawing on Black *et al.* (2011b), it is important not to overgeneralize the environmental impacts of climate change, let alone its effects on migration. Even the popular perception that climate change will lead to widespread drought in the countries bordering the Sahara is subject to question: 'Indeed, perhaps the only robust prediction that is currently available in relation to Sahelian rainfall is its likely continued geographic variability, such that even in the face of relatively extreme drought, some areas are significantly more affected than others' (Black *et al.* 2011b: 446). This is not to say that the impacts will be less or more than is sometimes predicted, for 'uncertainty cuts both ways' (Black *et al.* 2011b: 446). This variability matters: an uneven distribution of climate change impacts within the region is more likely to facilitate intraregional migration to neighbouring countries, while more homogenous negative climatic effects increase the propensity for longer distance migration. Therefore, not only are the impacts on migration uncertain, but they are also far from being universally 'democratic' even at the macro-regional scale, let alone the global.

MIGRATION AND HEALTH RISKS

Migration-Transmitted Health Risks in World History

Each human being carries millions of 'fellow travellers'—bacteria and viruses—in his/her body. These are mostly completely harmless, but some are deadly for other humans. Migration-transmitted diseases have resulted in death tolls of hundreds of millions and have been highly disruptive to many societies and civilisations. Migration provides bridges between health environments. The more diverse these environments are, the greater the chance of transmitting diseases for which a host country population has low or no immune response. The stereotypical image of an 'unclean or contagious foreigner spreading infectious diseases' has its roots in past narratives about the outbreak of devastating infections following contacts by previously isolated populations. In reality, the migration-enabled spread of infectious diseases has complex patterns, dependent on the type of diseases, the character of the affected population and the type of migrant transmitter.

The most fatal consequences of migration-transferred diseases were seen during the Age of Discovery in what Europeans termed the 'New World'. The effects on the population history of the Americas were especially devastating in the period 1492–1800, when European diseases devastated the populations of pre-Columbian societies. The Columbian exchange was so deadly because migration provided a bridge between previously unrelated health environments. Russell Thornton (1987: 42) estimated the total population of the 'Western Hemisphere' to be more than 72 million when Columbus arrived in 1492. Within a few centuries this had declined to an estimated 4–4.5 million due to the combined effects of diseases, alcoholism, genocide and forced relocations of tribes and nations.

The introduction of Eurasian diseases was lethal: 'The main killers were Old World germs to which Indians had never been exposed, and against which they therefore had neither genetic resistance. Smallpox, measles, influenza, and typhus competed for top rank among the killers' (Diamond 1997: 211). Weak immunity to foreign diseases by American Indians may have had many causes. Jarred Diamond (1997), for example, suggested that Europeans had a much wider variety of domestic animals than the indigenous population of America, and this cohabitation with domestic animals provided more opportunity for viruses and bacteria to mutate and migrate to the human population. Europe also was more densely populated and accustomed to exchanging waves of infectious diseases with Asia. European populations therefore had more frequent contacts with pathogens and higher resistance rates to the most deadly diseases, for which the indigenous populations of the Americas almost completely lacked immunity.

Some infectious diseases have travelled from the Americas to Europe. An outbreak of syphilis occurred in 1494–1495 in Naples, soon after Columbus's return from the Americas in 1493. Research on skeletal evidence of syphilis at the site in the Dominican Republic where Columbus landed suggested the route by which it was transmitted to the Old World (Rothschild 2005). This exchange of infectious diseases was a part of the famous 'Columbian Exchange', which included rapid and large-scale exchanges of people, ideas, flora, fauna, diseases, and cultural habits between the Americas on the one hand, and Europe and Asia on the other hand (Nunn and Qian 2010).

International migration also contributes to the spreading of infectious diseases in the early twenty-first century, albeit at much lower rates than in the Age of Exploration. Globally, tuberculosis, malaria and HIV/AIDS remain major infectious diseases that are transmitted via international migration (Gushulak and MacPherson 2004). The transmission of infectious diseases is assisted by the development of new transport modes, booming tourism and business travel, and the removal of travel and visa barriers in many regions of the world (Soto 2009).

Patterns of Migration-Related Health Risks

Some diseases have distinctive transmission patterns and relate to specific population groups. The diffusion of several sexually transmitted diseases (STDs, e.g. gonorrhoea, syphilis, viral hepatitis, herpes and chlamydia) was originally fuelled by military actions. The first symptoms of syphilis were observed in members of Columbus's crew. Phylogenetic analysis of the bacterium *Treponema pallidum* provides support for the Columbian theory of syphilis's origin (Harper *et al.* 2008). The disease reached Naples in 1493. It was quickly transmitted by French, Italian and Spanish mercenaries of the French king, Charles VIII, from Naples to the rest of Italy, France and Germany in 1494–1495. By 1502 it had spread as far as Denmark and Scotland (Quétel 1992: 10–11). While sailors, conquistadors and mercenaries were the original transmitters of syphilis, the disease quickly affected all population groups, with resulting pandemics. In the contemporary world there are some 500 million new causes of STD every year (WHO 2013a). The highest prevalence of the STD is in Africa, South America and the former Soviet Union. The transmission channels have shifted from soldiers and sailors to labour migrants, tourists and travellers, but the outcomes are similar.

Since the 1980s, HIV has become the most fatal global sexually-transmitted disease. In Europe, the spread of HIV seems to have coincided with major tourism, migration and labour migration routes. Paraskevis *et al.* (2009) found that Greece, Portugal, Serbia and Spain were major source countries, while Austria, Poland and Luxembourg mostly seemed to be recipients of the disease (Box 8.3). In West Africa, sub-Saharan Africa and South Africa, the rapid spread of HIV was related to massive rural-urban migration, disruption of traditional rural families and cultural values, and increases in poverty and prostitution (Corno and de Walque 2012). The spread of HIV was speeded up by particular high-risk population groups (lorry drivers, miners, labour migrants and prostitutes). In Thailand, the spread of the disease was also fuelled by its infamous sex-industry (Quinn 1994).

Box 8.3   HIV-Spreading Routes in Europe

Paraskevis *et al.* (2009) examined the distribution patterns of HIV-1 subtype B, the most common form of HIV in Europe. The analysis used samples from 16 European countries and Israel to map the geographical pattern of HIV genetic information. The HIV samples were taken at a number of locations and enabled the tracking of major "exporters" and "importers" of the virus. The study found that Greece, Portugal, Serbia and Spain were the major countries of export, while Austria, Poland and Luxembourg accounted for no significant exports of the virus. The densest routes of HIV-1 subtype B dispersal corresponded with the most popular travel and tourism patterns in Europe. Greece's migratory patterns were dispersed to seven countries, and those for Spain and the Netherlands to five and six countries respectively.

Some countries accounted for significant bidirectional migration patterns: Denmark, Germany, Italy, Israel, Norway, the Netherlands, Sweden, Switzerland and the UK. The strongest bidirectional pattern was found for the Netherlands, which has very diverse immigration communities, relatively open borders as a Schengen member within the EU, major ports and airports, and attracts some high-risk individuals susceptible to HIV infection (e.g. foreign drug users).

Some HIV migration routes correspond to patterns of labour migration. Specifically, high levels of virus migration were found from Italy to Austria and Switzerland, and from Portugal to Luxembourg (about 13% of the Luxembourg population is of Portuguese origin). Denmark was a migratory destination for virus spreading from other Scandinavian countries and Spain. Paraskevis *et al.* (2009: 11) suggested that 'prevention measures should not only be directed towards national populations, but also towards migrants, travellers and tourists who are the major sources and targets of HIV dispersal'.

Source: Paraskevis *et al.* (2009)

In contrast, tuberculosis seems to have originated some 40,000 years ago in Africa. The disease has two major lineages. The spread of the first one (human tuberculosis) is linked to human migration out of Africa. The second one (animal tuberculosis) is linked to humans infecting their domestic animals some 13,000 years ago in the Fertile Crescent. The human pathogen lineage strongly expanded approximately 180 years ago, as a consequence of population explosion and the industrial revolution (Wirth *et al.* 2008). Human tuberculosis originally affected all classes of the population but, in the modern era, tuberculosis is associated with poverty and lack of prevention. The World Health Organisation (2007) estimated that, in the early nineteenth century, 25% of deaths in Western Europe were attributable to tuberculosis, but the incidence of the diseases fell dramatically in the 1950s. The prevalence of tuberculosis, however, remains high in many less developed countries. Linked to this, immigration accounts for some 20% to 70% of the total number of cases notified in European countries. The incidence of tuberculosis among foreign-born populations living in Europe is up to 50 times higher than that of the indigenous populations (WHO 2007).

Why do (some) migrants have a relatively high propensity for transferring infections? Millions of migrants and refugees arrive from less developed and/ or conflict-ridden areas, where health care systems perform poorly. The migration process is also traumatizing. Many migrants have limited or no access to health assistance during periods of, often prolonged, movement (see discussion of fragmented migration in Chapter Seven). After arriving in the host countries, the migrants may experience a range of economic and social difficulties. They often face poor living and work conditions, cultural and language barriers, and prejudices in the majority ethnic group. All these factors amplify the risks related to the spread of a number of infectious diseases.

The Impacts of Migration-Transmitted Diseases on Societies

Some recent studies (Harbeck *et al.* 2013) indicate that the Plague of Justinian (AD 541–542) and the Black Death (1348–1350) were caused by the same pathogen (*Yersinia pestis*). Both plagues originated in China and spread via long-distance trade routes. The Justinian plague spread along sea trade routes to Constantinople. The Black Death spread along the Silk Road routes and sea routes from the Black Sea to Constantinople, and beyond to Europe.

The Justinian plague and the Black Death may have had the same origin and used similar dispersion routes, but they had markedly different impacts. The Justinian plague is estimated to have killed almost half of Europe's population. The plague is thought to have been a major factor behind the long decline of the Eastern Roman Empire, from which the empire never recovered. In contrast, the Black Death supposedly killed about a third of Europe's population, but did not cause the collapse of medieval Europe. In fact, huge losses of urban population hurt the aristocracy and clergy, who had sold agricultural surpluses to urban markets. The deaths of millions of people actually stimulated a dramatic growth in wages. This gave the lower strata of society some chance of improving their conditions of employment and living standards. Changes in the social and economic structure of medieval society eventually led to the establishment of national states and a capitalist production system.

Why were the impacts of the two plagues so different? Joseph Tainter, author of *The Collapse of Complex Societies* (1988), attributes the devastating impacts of the Justinian Plague to the complexity and hierarchical character of the Eastern Roman Empire. The Empire had a substantial urban population, which depended on grain supplies, tax collection and military recruits from rural areas. Unlike medieval Europe, with its lower shares of urban population, the Justinian Empire was more vulnerable to its economic and social fabric being disrupted by the plague. With high death tolls in rural areas, supplies of grain and taxes declined, and the Empire lacked resources to maintain its military forces. Attacks by 'barbarian tribes' further depleted the Empire's agriculture and tax resources, contributing to a vicious circle of decline (Tainter 1988: 145).

There are different views of the role of complexity in societal collapse. One view assumes that, in highly complex societies, hierarchical decision making gives way to distributed decision making. Decentralised decision-making systems create flatter hierarchies in organisations. The latter are more flexible and resilient than highly hierarchic systems. The loss of key individuals (perhaps due to migration-transmitted plague) does not necessarily cause the collapse of that society. Another view asserts that increased complexity leads to higher vulnerability. Shocks are more easily transmitted via tightly connected networks, and the impacts of failures are propagated rather than dampened in such complex networks. The more complex a society is, the greater the importance of key individuals, whose actions have significant implications for the rest of society. However, the evidence on the relationships between migration, complexity and societal collapse remain uncertain.

Management of Migration-Transmitted Health Risks

Are migration-related health risks more imminent in the modern World than in ancient societies? Globalisation, the removal of travel barriers and the introduction of new modes of travel enhance population mobility between different regions of the world. The World Tourism Organization, for example, has reported an increase in the numbers of international tourist arrivals from 436 million in 1990 to 1,035 million in 2012 (WTO 2013). Do global travel and migration create higher health risks? The higher intensity of exchanges both increases and decreases health risks. The more contacts there are between diverse environments, the higher the chance of transferring hitherto unknown diseases to host societies. At the same time, the risks of transmitting diseases between *radically different health environments* (such as the Columbian exchange) decrease. In the past, the populations of entire continents were susceptible to fatal invasions of new germs. Isolated populations are now rare. The functional bridges between diverse health environments are more numerous than ever before and so are the chances for transmitting infectious diseases. Yet the death tolls caused by infections have declined markedly since World War Two. This was, undoubtedly, influenced by large-scale vaccination programmes, the development of new antibacterial and antiviral drugs, and better cooperation between global and national health authorities. Some diseases have already (probably) been eradicated, such as polio. National and international infection control procedures have also suppressed outbreaks of some new highly pathogenic infections, such as SARS, Ebola and avian flu. The HIV remained the last deadly pandemic by the 2010s. Although it cost millions of lives, and caused untold misery, it did not destroy the fabric of societies.

Increasing flows of migrants, tourists and travellers need not necessarily increase global health risks, if they are properly understood, monitored and managed. The complexity of global mobilities, of course, requires a complex approach by migration policies. Particular mobility flows (tourism, business travel, labour migration, refugee movements, foreign aid) are managed by separate national and international bodies. These bodies pursue different goals and apply different (often incompatible) policy actions. Global health and migration governance has to recognise that migration is not a one-time and one-direction movement by an individual. Contemporary migration movements are typically characterised by the increasing importance of temporary and circular flows (Baláž *et al.* 2004). Any migration movement is usually a multistage process, which may include a pre-departure stage, travel stage, destination stage, interception stage (particularly for forced migrants) and return phase (Zimmerman *et al.* 2011). Health risks borne by specific migrant groups may be intercepted at different migration stages. Refugees, for example, tended to be attended to in the interception phase, labour migrants in host countries, while business travellers after returning to their home countries. Each stage and mobility group is the domain of specific national and/or international bodies. Some diseases develop quickly and are easy to recognise. Some traditional migration health policy instruments (such as quarantines and other frontier-based procedures) may be effective for controlling fast-developing infections (e.g. influenza, plague, SARS). They are less effective for diseases with longer periods of latency (e.g. HIV or tuberculosis). National and international migration and health policies have to recognise the diversity of mobility flows and migration-transmitted diseases. Information exchange and coordinated responses to migration-related health risks are the key to managing migration-related health risks.

Global disease surveillance and response was significantly assisted by new forms of international cooperation in health policies, and employment of new technologies. The Global Public Health Intelligence Network (GPHIN), for example, is an electronic public health early warning system developed by Canada's Public Health Agency. It continuously scans global media and websites for information on occurrences of infectious diseases and other international health risks. It is part of the World Health Organization's (WHO) Global Outbreak Alert and Response Network (GOARN), which combines the technical and operational resources of the WHO Member Countries and selected international bodies, such as Red Cross, UNICEF, and UNHCR (WHO 2013b).

INDIVIDUALIZATION, EDGEWORK AND ADVENTURE-SEEKING MIGRATION

At the heart of the concept of individualization is the notion that, in the risk society, there has been a weakening of the old certainties, and the values attached to class and estate-specific knowledge. This has been replaced by greater uncertainties and detraditionalization, as a result of which there has been 'a social surge of individualization' (Beck 1992: 87). The individualization thesis (Beck and Beck-Gernsheim 2002) sees individuals as taking increased responsibility for their life course trajectories, and for their careers. In line with the notion of reflexive modernism (Giddens 1991: 135), biography has become less a matter of ascription, and more of a reflexive project that is 'self-produced and continues to be produced' by individuals. This resonates with Rose's (1990: 15) notion that individuals increasingly display resourcefulness, driven by 'motives of self-fulfillment'.

Beck's and Gidden's ideas about individualization and reflexivity have been extended, and in some ways been made more concrete, by Lyng's (2008) intermediate concept of edgework. This explicitly recognises the 'seductive power of the risk experience' (Lyng 2008: 120), that is, the positive evaluations of risk:

> Confronting and responding to uncertainty is what edgeworkers value most, even as they devote significant effort to managing risks in order to reduce the likelihood of hazardous outcomes.
>
> (Lyng 2008: 109)

Edgeworking suggests proactive responses to confront risks, rather than passive acceptance or simply responding to risks as these are manifested. It also recognises that confronting risk is a form of 'work' that is valued by individuals. The 'work', in the term edgework, is about developing personal competences in managing risk and uncertainty, and in this limited sense can be related to the behavioural economics of willingness to take risks and competence to manage risks (Chapter Three). Edgework is exemplified by growing preferences for riskier lifestyles, whether at play or at work, but also—as discussed below—in mobility and migration.

Most forms of migration involve an economic calculus, sociocultural motivations, or some form of economic, political or environmental 'push'. The end object is realised in the destination, and migration is seen as involving risks both negative and positive. However, some types of migration positively value confronting the risks in the migration process itself. This is typified by young adult mobilities and temporary migrations, ranging from, say, adventure tourism to backpacking, to student migration, to temporary labour migrations, or what is known as 'the Big OE' (Overseas Experience) in Australasia. These all represent edgework—opportunities for 'successfully' engaging with risks and uncertainties. The participants positively value the risks and uncertainties associated with migration, above all the opportunities to 'work' on these, and thereby develop competences to manage and overcome the risks. Such experiences are sources of enhanced self-esteem, and exemplify reflective modernity in practice. They also contribute to biography creation, which is most clearly exemplified by how such international migration experiences contribute to CVs, and to strengthening individuals' labour market competitiveness. Temporary migrations, and the opportunities for biography creation, have been facilitated by transport technology changes, and organisational innovations related to low-cost carriers, which have reduced both the accessibility and cost barriers to international travel (Williams and Baláž 2009).

Student migration can be partly interpreted in terms of the individualization thesis, according to both King and Ruiz-Gelices (2003: 232) and Baláž and Williams (2004). They present case studies of shorter-term placements of UK and Slovak student migration, respectively, that is, excluding those students who spend three or four years studying abroad for an undergraduate degree or doctoral thesis (Box 8.4). A central argument in both studies is that student exchange schemes, such as ERASMUS, are a means of biography construction. Individuals can significantly enhance their CVs compared to those who have not studied abroad, because their 'successful' migration experiences provide evidence of a range of competences, including confronting and managing risk and uncertainty. In other words, mobility experiences contribute to 'do it yourself' biography.

Box 8.4   Student Migration, Do-It-Yourself Biographies, and Individualization

Studies by King and Ruiz-Gelices (2003) and Baláž and Williams (2004) considered somewhat different student migrant groups, although both spent relatively short periods abroad, mostly being 3–12 months. The King and Ruiz-Gelices study analysed the experiences abroad of a cohort of students from the University of Sussex (UK), who had spent a year of their four-year degree programme studying abroad. Their main data source was 261 questionnaires completed by these students, although they also undertook supplementary surveys of students who had not studied abroad, and those about to depart on a year abroad. In contrast, the Baláž and Williams (2004) study was based on a sample of 55 in-depth interviews with Slovakian students who had returned from a period of study abroad of between 3 and 12 months. Just over two thirds of their sample had studied abroad as part of their degree programme, while just under a third had been on language courses or, in a few instances, vocational courses.

The Sussex sample were aware of the negative as well as the positive risks of spending a year abroad, and only 13% considered that they did not have any worries or concerns before they left Sussex. Similar pre-migration information was not available for the Slovak sample.

The evaluations of the period abroad in the two studies were both very positive (Table 8.1). In the Sussex study, 90% of respondents considered they were satisfied or very satisfied with the outcomes. Of particular relevance for this study are the evaluations of specific aspects of their personal development, or competences. The Sussex students saw the main benefits in terms of personal development (which seem to equate to positive outcomes of acquiring competences via edgeworking) and acquisition of language skills, while obtaining knowledge of a foreign country and new perspectives on their own country were also important. The contributions of migration to their general career prospects via biography building were also seen as positive overall. Since

Table 8.1   Returned student migrants' evaluations of the acquisition of competences and personal development

a) Sussex sample

	Ranking (1 = maximum satisfaction (5 = minimum satisfaction)
General enhancement of academic and professional knowledge	2.5
Relevance to current job	2.8
Relevance to general career prospects	2.3
Foreign language competency	1.7
Maturity and personal development	1.6
Knowledge and understanding of foreign country	1.8
New perspectives on home country	2.0

Source: after King and Ruiz-Gelices (2003: 237)

b) Slovakia sample

	Ranking (1 = maximum satisfaction (5 = minimum satisfaction)
Acquired qualifications	3.2
Learned new skills	2.8
Acquired new ideas	2.0
Better able to deal with challenges	2.0
Learned new approaches to work	2.0
Foreign language competency	1.4
Enhanced self-confidence	1.8

Source: after Baláž and Williams (2004: 225)

graduation, they were more likely to have achieved higher level jobs, higher salaries and to have migrated, compared to students who had not had a year abroad.

Although the questions discussed were somewhat different in the study of Slovak students, the results are broadly similar, with foreign language competence and personal development, in the form of enhanced self-confidence, being highly rated. Ranked just behind these was the acquisition of other forms of personal developments or competences, including being better able to deal with challenges. The last, in particular, indicates the importance of acquiring competences to manage risk through edgeworking. While no comparative study of non-migrants is available to evaluate how competences and biography building influence their career prospects, we can note that just over one half of the Slovak sample recorded an improvement in their social status, employment status or income; given that many were still completing their studies, it can be anticipated that these figures would rise over time, after they had graduated.

The two studies draw broadly similar conclusions about the relevance of the individualization thesis. King and Ruiz-Gelices (2003) provide the stronger endorsement of this theoretical perspective: 'Students who choose to study abroad are taking a significant step in setting in motion their own individualised life-projects. . . . They start to create their own "bricolage-biographies" . . . with links potentially to several places, skills, languages and cultures' (King and Ruiz-Gelices 2003: 246–247). Baláž and Williams (2004) consider that their findings are broadly in line with those of the Sussex study, although they also underline the importance of the structural determinants of individual experiences. They also emphasise the need to place the migration experience in context of the full migration cycle or cycles, or the life course. Many returnees—for various reasons including positive experiences, the networks they have acquired, or greater competence to manage the risks associated with migration—have a relatively high propensity to remigrate. King and Ruiz-Gelices (2003: 232) also comment that: 'The YA [Year Abroad] may be the first step in constructing an "intercultural lifeworld", which becomes more intense if the foreign residence is prolonged after graduation or if cross national partnership or marriage takes place'.

Sources: King and Ruiz-Gelices (2003) and Baláž and Williams (2004)

The individualization theory offers insights into understanding specific types of migration, such as student migration, not least in making links with broader theories of social change. However, it is also problematic. One of the central tenets of Beck's (1992: 127–8) argument is the existence of fundamental shifts in how individuals are embedded in the risk society. He contends there has been: 'disembedding from . . . historically prescribed social forms and commitments', and processes of re-embedding whereby individualization becomes relatively more important, as traditional institutional forms decline in importance. In contrast, Abbot *et al.*'s (2006) meta-review of risk and security issues in the labour market provides both confirmation and qualification of the individualization thesis. They conclude that individuals may become more reflexive about risks, but that reflexivity and the distribution of risk and insecurity are 'highly dependent on socio-economic factors and existing inequalities' (Abbot *et al.* 2006: 240).

Similarly, although Baláž and Williams (2004) and Williams and Baláž (2004) found evidence for reflexivity and taking responsibility for biographies amongst, respectively, students (see Box 8.4) and au pairs, they also argued for the continuing importance of structural considerations. Amongst those who had returned to jobs in Slovakia rather than further study, there were real differences between the more positive employment experiences of those who worked in the private sector, especially the international segment, and the public sector. The latter were far more likely to consider that their experiences were undervalued. Edgeworking does not always result in material or occupational returns, even if the individuals still benefit—at least initially—from self-esteem. This is consistent with Tulloch and Lupton's (2003) argument, noted earlier, that the complexity of risk and uncertainty behaviour makes it difficult to analyse in terms of the changes in reflexivity and individualization. Lash (2000) similarly has argued that complexity, contradictions and ambiguity in most aspects of everyday life are more important than is suggested by the risk society thesis. Finally, in an even more strident critique of the individualization thesis, Zinn (2008) argues that in some ways it romanticises how individuals, by assuming greater responsibility in biography building, are seemingly liberated from traditional social ties. It ignores the way in which individualization generates new forms of risk because, especially in the face of weakened institutional ties, 'the elective biography can all too easily become "the breakdown biography"' (Zinn 2008: 33).

CONCLUSIONS: MIGRATION AND RISK IN CONTEXT OF SOCIETAL AND TECHNOLOGICAL CHANGES

The relationship between risk and migration is dynamic and is mediated by long-term technological changes in societies. This is a long established process which has seen migration being both generated by, and generating, risks associated with major social and technological shifts. However, writers such as Giddens have argued that there has a been a shift to a more reflexive modernism in recent decades, and one of the most prominent theorisations of how this mediates understandings of risks has been Beck's (1992) risk society thesis. In essence this argues that societies increasingly are characterised not by too little knowledge, but by too much knowledge, and a failure to be able to control this. The consequences have become 'democratised', with income and other social distinctions no longer being a moderator of the failure to manage such risks.

These broader social and technological shifts do play a role in shaping the relationship between risk and migration. However, the risk society theory is unable to engage effectively with the complexities of migration. This is evident in the examples provided in this chapter of both health risks and environmental refugees. The latter provides one of most contentious and yet clearest illustrations of the risk society thesis. Climate change has its origins in technological changes dating back to the first industrial revolution, so that many current migration flows are in fact the cumulative outcomes of much earlier periods of technological changes. Moreover, large-scale environmentally-informed migration does not follow in any simple way from climate change, even in terms of aggregate outcomes, because a number of responses are possible including adaptation. Transport technology shifts have had major health consequences, but these have been evident from at least the classical period, when the growth of shipping and trade constituted strong risks of potentially devastating epidemiological outcomes. And with both health and environmentally induced migration, there is no simple 'democratisation' of consequences: wealth, power and resources continue to be important mediators of who can adapt or who can flee in advance of, or in the aftermath of, major failures to control knowledge and technology.

9 Migration Discourses in the Face of Risk

INTRODUCTION

This chapter considers the constructionist approaches to risk, and in particular the production and influence of discourses about migration and risk. As Masuda and Garvin (2006: 437) argue, 'What constitutes danger (or negative risks) depends on 'who is talking to whom''. Whether or not risks have existence 'out there' beyond discourses is, for our purposes, less important than understanding those discourses. At one level, this is about individuals' perceptions and understandings, but these are not the central focus of this chapter. Instead, in both this chapter and in Chapter Ten, the focus is on understanding discourses on migration and risk in context of a much broader societal framework relating to the relationships between risk, governmentality and governance. This is because risk—since at least the emergence of modernism—is 'a way in which we govern and are governed' (Adam and van Loon 2000: 2).

The cultural theorists, encountered earlier in Chapter Six, have addressed how particular groups, in some way at the margins of society, are identified with risks to the social order. Douglas (1992: 7), for example, writes that 'certain marginalized groups are identified as posing risks to the mainstream community, acting as the repository for fears not simply about risk but about the breakdown of social order and the need to maintain social boundaries and divisions'. The identification of which groups are a risk, or at risk—and both categories apply to migrants—is not produced randomly, but is infused with power relationships. The media are especially powerful agencies in generating and distributing discourses, being driven by competing interests both within the media organizations themselves, and broader agencies in the wider society. These other agencies include academics, NGOs, businesses and their associations, and the various components of the state and quasi-state. Even this book, in its own modest way, contributes to the production of such discourses. There are a number of theoretical approaches to understanding discourses, but Foucault's governmentality argument is probably the best known, and is considered later in this chapter (see also Foucault 2007).

The focus on discourses implies more than just the identification of groups posing risks to the social order. It is also about the language that these discourses are expressed in. As Papademetriou and Heuser (2009: 21) comment, 'words and meanings can become imbued with new meanings, depending on how and in what context they are communicated'. This is because risk is performative, and ultimately such performances produce the effects that they indicate (Butler 1990). Headlines in the popular press about 'illegal migrants' or 'migrant cultures' are likely to use far more emotive language than, say, the *Economist* about migration, but language is not neutral in either of these. Rather it is formative and imbues meaning to ideas and processes. In the former, for example, the cost-benefit analysis of the contribution of migrants to a national economy is reduced to the risks of migrants being 'job stealers' or 'burdens on the public health system'.

Discourses about migration and risk are strongly and persistently developed, polarised and polarising, and contested. For Masuda and Garvin (2006: 437), this is because migration, as with most issues, becomes situated at the juxtaposition of the communication of 'contested ways of making sense of the world': different groups advance their own views of risk, and these are deeply reflective of their world views. And, as already emphasised, it is not just the ideas about migration that are important because 'the language used to talk about immigration and immigrant integration does matter (Papademetriou and Heuser 2009: 19). In the UK, the Member of Parliament Enoch Powell's 1968 'river of blood speech' is perhaps one of the most famous, or infamous, examples of how ideologically powerful and emotive language and metaphors have shaped discourses and influenced policies (Hansen 2000). This is a particularly powerful example of both the creation of discourse through oratory and its amplification and transformation by the media, because the phrase 'rivers of blood' never appeared in the actual speech, but instead refers to a line from Virgil about 'the River Tiber foaming with much blood'. It is an exemplar that subsequently has echoed down the years because, as Threadgold (2009: 223) notes, 'the media have covered asylum and refugees using a template or frame, that invokes "floods", "invasions", "criminality" and government loss of control'. These are strongly recurrent themes for, as Berkeley *et al.* (2006: 28) comment, migration discourse 'does not disappear, but resumes in new forms and configurations'.

The most strongly developed discourses tend to centre on who or what is at risk, whether amongst particular groups of migrants or in terms of real or imagined aspects of the shared ways of life of particular communities. These are hardly new, and discourses about migrants as 'threats' have echoed down the centuries. However, Papademetriou and Heuser (2009: 22) consider that such discourses have intensified in recent decades: 'Broadly speaking, public anxiety about immigrants and immigration has increased across the globe in the past decade'. Moreover, the sheer pace of change contributes to the ability of discourses to shape, and resonate with, growing anxieties about risks to popular understandings of shared ways of life. That begs the question, of course, of the extent to which there are shared ways of life in societies which are often riven with class, gender and ethnic cleavages, but such issues are sidestepped rather than confronted by most discourses about migration.

Much of the academic literature in this area focusses on security rather than risk per se (Rudolph 2003), and the globalisation of migration is often presented as a discourse on 'internal security' and threats to stable national identities and particular ways of life. This has become particularly marked in the post-Cold War period although, as noted, the themes are persistent and have historical parallels. Ceyhan and Tsoukala (2002: 22) comment that a prominent feature of what they term the 'postbipolar era' in the West has been:

> The production of a discourse of fear and proliferation of dangers with reference to the scenarios of chaos, disorder, and clash of civilizations. It is easily noticeable in the public sphere that the fear is mainly about the different, the alien, the undocumented migrant, the refugee, the Muslim, the 'non-European,' the 'Hispanic.' These different expressions converge on the figure of the migrant, which appears as the anchoring point of securitarian policies and fierce public debates.

These discourses feed 'a new political imagination preoccupied with the play of mobilities, and populated by elusive persons (terrorists, asylum seekers, smugglers) and mercurial things (contraband, drugs, weapons) that are able to move about almost undetected, exploiting the smooth, networked spaces, but also the seemingly ungoverned borderlands of a "global world"' (Walters 2008: 170). In this world view, risk resides in marginalised groups—'these elusive persons'—in marginal spaces of the global world, but presented as being situated uncomfortably close to, and encroaching on, a supposedly stable, harmonious and—frequently—homogenous society built on shared beliefs and ways of life.

The academic literature on migration discourses in the West has mainly focussed on these different expressions of the *im*migrant, and far less explicit attention has been given to discourses about emigration and risk in the countries of origin. The persistent and tragic deaths of migrants seeking to cross the Mediterranean, or from West Africa to Spain's Canary Islands, has, however, led to some researchers seeking to understand how various discourses have sustained these migrations, in the face of well-reported risks to life. With both immigration and emigration, the issues at the heart of these discourses about risk are not necessarily new. Rather, as Amoore and de Goede (2008: 9) comment in relation to the 'war on terror': 'It is not strictly the case that new risks have come into being but that society has come to understand itself and its problems in terms of risk management'. Popular and policy discourses are central to this understanding.

The remainder of this chapter considers three main themes relating to discourses about migration and risk. First, we consider the nature of these discourses, and how they are produced, especially in the media, drawing on notions of governmentality. Second, we turn to the selective portrayal of migrants as being at risk, and specifically as being victims. Finally, we consider some of the prevalent discourses about migrants being outsiders, and their social construction as being sources of risk.

GOVERNMENTALITY, DISCOURSES AND THE MEDIA

An approach to discourses which has particular leverage in migration studies is governmentality (Williams and Baláž 2012). Here we draw on Foucault's work, while noting that there are other understandings of governmentality (see Dean 1999). Foucault (1991) focusses on the practices rather than the institutions of governments, and one of his central arguments is that power is widely distributed across society through various practices and discourses that produce knowledge, rather than being concentrated in government.

As will be discussed in the next chapter, the emergence of modern states was associated with an increasing focus on risk management—whether this was in relation to economic or demographic issues (Zinn 2004a). However, as we have emphasised, it is not so much actual events but discourses about these which are critical. As Zinn (2007) comments, 'It is not a specific event that constitutes a risk, but its *description* as part of a risk calculation make it a risk' (Zinn 2007: 17, emphasis added). These descriptions of groups, considered to be either at risk or sources of risk, are powerful because 'generalized social categories in institutional and media discourses produce homogenous groups in relation to risk' (Zinn 2004a: 12), whereas such groups are usually highly diverse.

In migration, this production of generalised social categories is especially evident in relation to groups such as refugees, illegal or irregular migrants, and those who are trafficked.

Governments do not sit outside the arena in which discourses are produced, but rather they play a role in their production. For example, Edwards (2009: 9) contends that both the EU Commission's first political assessment of progress on the implementation of the Hague Programme and the UN's National Convention against Transnational Organized Crime serve to assemble together, and often to merge, crime, terror and immigration. This contributed to the creation of 'a culture of fear'. Governments are not the only actors involved in creating and sustaining such social categorisation and a culture of fear, because—as Foucault argues—power is distributed across society. It also resides in NGOs, in local community groups, and amongst migrants themselves, to varying extents. They all play roles in defining the risks associated with migrants, or in categorising 'risky migrants'. Whether they reinforce or challenge the dominant discourses, they thereby simultaneously define and produce these risks. The media—itself a very diverse set of actors—is probably the most powerful of the actors which contribute to discourses about migration and risk.

Press coverage of migration has always tended to peak around what editors and journalists consider to be 'newsworthy' events. This is necessarily highly selective, both in respect of the 'events' or features covered, and how they are presented. The cumulative effect is the creation of, or contribution to, powerful and highly influential discourses about migrants and migration. Suro (2009: 195) contends that while this pattern can be traced back over many decades, 'the fluctuations in the volume of coverage have become even wilder in recent years'. Indeed, Aoki (1996: 51) argues that 'with the rise and global spread of powerful and pervasive film, television, and culture industries in the twentieth century, this re-imagination has accelerated and become pervasive'. Moreover, the transformation of the media which has resulted from the growth of cable and satellite television, and the Internet, have reinforced traditional journalistic practices, thereby tending to magnify these cumulative distortions in the coverage of migration (Suro 2009). The influence of the Internet, however, is more complex in that the social media facilitate the production and distribution of a wider array of discourses.

Suro (2009: 201) contends that the nature of migration means that it lends itself to the production of discourses by the media.

> All storytelling, whether factual or fictional, is easiest when narratives can be constructed around the actions of a single person or a group of people. Narratives beg for protagonists, whether they are heroes or villains, victims or perpetrators. This imperative can have particularly perilous consequences when applied to a phenomenon like immigration, and yet migration lends itself to simple narratives in which the migrant is the obvious protagonist. After all, moving from one country to another provides a clear plot with a beginning, middle and end: It is the kind of dramatic action which readily drives narratives, especially when it involves physical peril, actors of illegality or both.

The stories told in the western media about migrants are diverse, but in aggregate they tend to 'portray members of migrant and minority populations in the worst possible light: as criminals, terrorists, and more generally as people who represent a threat to the established way of life' (Boucher 2008: 1465). While this may be an overstatement, and one that disregards the diversity of the media, the influence of the latter is undeniable, as is illustrated by the examples of the UK, Germany and the USA.

In the UK, Saran (2009: 156), drawing on the work of the IPSOS Mori polling agency, concludes that popular British anxieties towards immigration are linked to both increased immigration and to greater media focus—which resonates with Suro's (2009) argument that the focus on what is considered newsworthy has become wilder in recent years. Of particular note is the importance of uncertainty as opposed to risk in such discourses: 'The current debate is often framed in such a way that immigration is often presented as a phenomenon which has led to rapid unexpected changes and widespread uncertainty' (Saran 2009: 156). And, as we have noted in the work of the behavioural economists (Chapter Three), individuals tend be more uncertainty-averse than risk-averse, which magnifies the power and reach of such discourses.

Turning to Germany, Zambonini (2009: 175) similarly notes how, in the 1980s and 1990s, asylum became a major issue in German society. The German media played an important role in shaping emerging debates about migration, tending to focus especially on immigrant crime, and frequently presenting migrants as being asylum seekers peddling false claims. '*Das Boot ist voll*' (the boat is full) was a slogan that seemed to echo through contemporary public discussion, and often appeared on the banner head-lines of some of Germany's best known news magazines such as *Der Spiegel* (Zambonini 2009: 169).

Finally, a similar picture of the role of the media in the production of discourses is evident in the USA. Suro (2009: 197) comments that over a prolonged period of several decades, American journalists and editors have tended to focus on the illegality or the uncontrolled nature of migration while largely ignoring the significantly larger, orderly, and regular migrations to the USA. In other words, the focus has been on the risk to social order. In the period 1980–2007, there were 1,847 Associated Press stories on immigration, and almost four out of five of these focussed on illegality, which is vastly disproportionate to the number of illegal migrants. However, as in any western country, the press coverage in the USA has been diverse, with key individuals or agencies being highly influential in the development of such discourses. Suro (2009: 199) focusses especially on the role of CNN's Lou Dobbs who:

> led the way in characterizing unauthorized immigrants as threats to the health and safety of ordinary Americans, portraying them as a category of people who are not merely undesirable but who need to be expelled in order to preserve the nation . . . He has frequently used the language of conquest, an 'army of invaders' to describe the migrants.

Since the events of 2001 a further twist has been given to this not only in the USA but in most, and possibly all, developed economies. Arab or Muslim identity has become a fulcrum of discourse about terror (Ibrahim 2005), a theme that we return to later in this chapter.

Of course, not all media stories about migration are negative, and there are also discourses which contest dominant stories about migrants as sources of negative risk and uncertainty. This was explicitly recognised by Foucault (1980: 101) who emphasised that it is important that we make allowance:

> for the complex and unstable process whereby discourse can be both instrument and effect of power, but also a hindrance, a stumbling-block, a point of resistance and a starting point for an opposing strategy. Discourse transmits and produces power; it reinforces it, but also undermines and exposes it, renders it fragile, and makes it possible to thwart it.

These counter discourses may originate from the migrants themselves, their associations, the NGOs which appoint themselves as their champions, or from individual journalists and editors—illustrating Foucault's argument about the production of discourses being dispersed across society. And, increasingly, they may also originate—as do more negative discourses—from the social media.

The media also play an important role in generating emigration, not least in the images they present of desirable forms of consumption, and stories about opportunities and success. The oldest of all of these is the mythical notion of the 'streets being paved with gold', illustrated in John Steinbeck's (1939) *The Grapes of Wrath* by the promise of well paid jobs for displaced Oklahoma farmers, picking fruit in California in the interwar Great Depression and dustbowl years. In reality, of course, these images and discourses interact in complex ways with diverse individual models of personhood, as Mai (2004) explains in the case of Albanian migrants to Italy (Box 9.1). Rudnyckyj (2004) brings an additional Foucauldian (1997) twist to understanding how the production of knowledge influences emigration. The author argues that technologies of servitude, produced by human resource organizations and NGOs in particular, impart desired (by prospective employers in destination countries) skills and attitudes to potential migrants. Such mundane technologies 'are intended to rationalize performance, profitability, and security' (Rudnyckyj 2004: 430).

Box 9.1   Albanian Migration to Italy: The Role of the Media

As the command and regulatory structures of the former state socialist societies of Eastern Europe frayed, and then collapsed, after 1989, a number of major migration flows emerged, one of the most dramatic and widely reported of these being the migration of young Albanians to Italy. Mai argues that these migrations have to be understood as being shaped at the nexus of media consumption, social change and the imagination and enactment of mobility. Moreover, he contends that young Albanians have been subject to competing and contradictory regimes of subjectification (see also Rose 1996b). Specifically, he contends that there was a meeting between very different, and largely contradictory, models of personhood: between the traditional family-centred narratives of Albania, and the highly individualised life trajectories that were portrayed in Italian television programmes. The outcome of this was the large-scale young adult migration that characterised the post state socialist period.

Interviews undertaken with newly arrived Albanian migrants indicated the particular role played by Italian television in the later years of state socialism. Almost all of the migrants (97%) stated that they had watched Italian television on a regular basis, while 89% considered that they had learned Italian in this way (Dorfles 1991: 14). This was a period in which Italian television had 'the status of the privileged "window on the world"' (Mai 2004: 7), and was the dominant shaper in Albania of discourses about life in Italy.

By the late 1990s the position had changed dramatically. The monopoly of Italian television as a source of discourses about the promised fruits of migration to Italy had been replaced by a greater diversity of sources of information including a range of television channels. More importantly, there were also 'the narrative accounts of disillusion with capitalism which were provided by returning migrants' (Mai 2004: 7). Despite these competing narratives, Mai's (2004: 18) own interviews—undertaken in 1998–2000—found that most of the young people interviewed were still willing to emigrate, and most recognised that this was influenced by watching foreign television. He concludes that 'both older and younger interviewees acknowledged the role of media in providing both information about potential destinations and, most of all, alternative lifestyles which are consistent with different ways to relate to and conceive the self in more individualist and fun-oriented terms'.

Source: Based on Mai (2004)

DISCOURSES ON MIGRATION AS A SOURCE OF RISK

Social problems are socially constructed in order to influence public opinion to be receptive to particular rationalizations (Edelman 1988). This form of social construction has been particularly marked in migration. Ceyhan and Tsoukala (2002: 23) argue that:

> Since the 1980s, migration has become the catalyst supposed to be able to summarize most of the current social problems of Western societies. By a sidestepping of the nonsecuritarian insights of economic, social, and cultural analyses, immigration is now apprehended under the nearly exclusive angle of securitarian and identitarian preoccupations.

In other words, the selective focus on social identity and security issues, to the exclusion of all others (such as cosmopolitanism, or economic contributions) has dominated popular understanding of immigration. Weiner (1995) provides one example of the view that increased international migration in recent decades has been seen to pose threats to international stability and security, most obviously in the Balkans and other fragile areas, but also within western democracies because of xenophobic and nationalistic politics. In this view, 'the most advanced industrial democracies risk being destabilized politically by a massive influx of unwanted immigrants, refugees and asylum seekers' (Weiner 1995: 188).

Ceyhan and Tsoukala (2002: 24) argue that discourses presenting migrants as social problems—for which we can also read sources of negative risk for the host society—centre around four main axes.

- A socioeconomic axis, which focusses especially on the threat to the employment of indigenous workers, and to the costs of the welfare state.
- A securitarian axis, where the risk of a loss of control over borders is seen as a risk to both security and sovereignty.
- An identitarian axis, wherein migrants are presented as threats to the national identity of the host society.
- A political axis, characterised by anti-immigrant and racist discourses.

The following discussion explores how these four—sometimes interwoven—discourses are illustrated in selected examples. Starting with the USA, Delgado and Stefancic (1992) contend there are remarkable similarities, in American history, in the racial stereotyping of discourses about successive migrant groups, whether Africans, Mexicans, or Asians, reflecting both white/nonwhite classifications, and Orientalist racial discourses. Aoki (1996: 24) writes that characteristics such as 'primitivism, savagery, foreignness, and unbridgeable racial difference presents' were ascribed to virtually all nonwhite migrants.

Chinese migrants, in particular, were presented as risks in terms of Ceyhan and Tsoukala's (2002) socioeconomic and identitarian axes. They were 'heathen, morally inferior, savage, childlike, and lustful' (Aoki 1996: 24). The language of the discourses about Chinese migrants in the nineteenth century foreshadowed that employed by Enoch Powell and others in the second half of the twentieth century, with stereotypes representing the risks to American society as 'different sorts of natural disasters: floods, tidal waves, inhuman swarms, and plagues' (Aoki 1996: 31). But Aoki also cautions against oversimplification, arguing that such racial stereotypes are both stable and fluid. Thus, with a small but significant change of emphasis, during World War II the Chinese ceased to be a 'yellow peril' and became a 'model minority', although a decade later—in the postwar context of the rise to power of Mao and the Communist Party in China—they had become the 'horde-like and godless Red Chinese menace' (p.16).

When the USA entered the Second World War, and with China portrayed as victims of Japanese aggression, the American Japanese replaced the Chinese as the principal target of hostility. The risks they posed were presented especially in terms of the securitarian axis. Aoki (1996) provides a vivid account of how this particular discourse of negative risks led to the internment of West Coast Japanese and Japanese Americans during World War II (see Box 9.2). Their treatment was very different to that of the Italians and Germans in the USA (Fox 1988). In the six months after the Axis powers declared war on the USA in December 1941, the American government relocated some 10,000 German and Italian aliens from the West Coast because of their perceived risks to America. The government also considered

---

Box 9.2   The Internment of Risky 'Migrants' in the USA during World War II

After America entered World War II, some 120,000 American Japanese living on the West Coast were hurriedly interned. This was against a background of strong popular antagonism against these populations which both the government and the media contributed to. Of course, anti-Japanese propaganda 'was not created out of thin air but worked with and amplified pre-existing and well-defined racial stereotypes that had deep historical roots' (p. 37).

'This was a striking and tragic example of wholesale inscription of loyalty to a foreign government and attribution of unproved and ultimately unfounded subversive intent against the United States' (p. 38). There was a classic process of othering, as American Orientalism was inscribed on the Japanese, and they became characterised as the 'yellow peril' in place of the Chinese. Powerful and emotive language was employed to describe them such as 'treacherous, sneaky, cruel, bloodthirsty, fanatical, suicidal, scheming, fatalistic, non-Christian' (p.38), all of which were considered as being the opposites of the supposed characteristics of the typical American citizen.

Perhaps the most telling comparison is with the treatment of the German and Italian populations, which were not interned en masse in the way that the Japanese were. The language of the relevant discourses was also different. The Japanese were described as 'Japs' or 'Nips', terms which had both racial and national undertones. In contrast, there was reference to the 'Nazis' in Germany, which broadly equated to the notion of 'bad' Germans, implying that there were also other—presumably decent—Germans. In other words, there was not the universal racial stereotyping and demonizing of Germans and Italians as evil, untrustworthy sources of risk to American security, as happened with the Japanese. 'Ethnicity, national allegiance, and race literally intersected on the site of the Japanese and Japanese American body, requiring swift incarceration, strict containment, and distant spatial segregation to preserve the integrity of the "American" body politic' (p. 39).

Akoi concludes that 'if every man, woman, and child of Japanese descent was a potential menace, then the actual extra-national Japanese enemy could be constructed as an even more incorrigible and monolithic nemesis, an enemy whose menace came both from *within* and *without* the nation-state' (p. 40).

Source: Aoki (1996)

interning them, along with tens of thousands of German American and Italian American citizens, but ultimately desisted from such a move. This poses the question of why only the smaller number of Japanese was interned en masse. Fox (1988: 409) considers the answer lies in part in racism and racist discourses, sensitivities about how German Americans had been persecuted during World War I, and fear of alienating literally millions of the American-born children of Italian and German parents and grandparents. German and Italian Americans also benefitted from the sheer numbers involved, invoking concerns about the practicalities of internment. They also had the advantage of having become more assimilated than the Japanese into American society. In contrast, the 'ineradicable foreign-ness' of the Japanese was part of the justification for their internment.

Moving forward a few decades, some of the most powerful migration discourses have focussed on asylum seekers and refugees. Although the term 'asylum seeker' is now well embedded in media, policy and academic discourse, it only began to be more widely used in the early 1980s (Collyer and de Haas 2010). Prior to this, there had been an almost universal agreement that refugees were at risk and needed protection, so that individuals were categorised as refugees from the onset. In contrast, in the final decades of the century, the claim to being at risk was not automatically assumed, and indeed was presumed to be without foundation until 'proven' otherwise (notoriously difficult given the circumstances in their countries of origin). In the interim period, individuals were described as 'asylum seekers' and for some elements of the British tabloid press, for example, this became 'a shorthand for undeserving and fraudulent' and shaped the coining of a common vocabulary which, in due course, served to undermine much of the existing de facto and de jure protection provided for refugees in the UK (Collyer and de Haas 2010: 472). Similar recasting of the vocabularies of discourses about migration can also be observed in other countries.

While the discourses about asylum seeking were initially positioned on the socioeconomic axis, since the 2001 attack on the World Trade Centre in New York, discourses about migration have shifted to the securitarian axis in most developed countries. Huntingdon (2004) (in)famously argued that in the post-Cold War era, failure to control American borders was the single biggest threat to the national security and identity of that country, in context of both globalisation and multiculturalism.

The securitarian axis was not, of course, an entirely new influence, as evidenced by the examples of internment during World War II. Moreover, immigration was increasingly presented in terms of the negative risks it posed to public order and social stability from the 1980s (Huysmans 2000). In Europe, the ending of the Cold War, the growth in international migration, and the formation of the EU's Schengen Area all contributed to the securitisation of migration discourses (Ceyhan and Tsoukala 2002: 21). Since the 2001 World Trade Centre attacks, questions relating to migration and security are increasingly viewed through the lens of international terrorism in Europe as well as in the United States. Foreign minister Josep Piqué of Spain, for example, argued that the 'fight against illegal immigration is also the reinforcement of the fight against terrorism'. A recent report published by the Nixon Center declares, 'Immigration and terrorism are linked—not because all immigrants are terrorists but because all, or nearly all, terrorists in the West have been immigrants'. The same report goes on to cite Rohan Gunaratna's claim that 'all major terrorist attacks conducted in the last decade in North America and Western Europe, with the exception of Oklahoma City, have utilized migrants'. Such claims are sensationalist and highly problematic, not least because they do not take into account attacks by domestic groups in Europe such as the separatist group Basque Fatherland and Liberty, also known as ETA. Also, many *jihadists* have been born in relatively long-established ethnic minority communities in western countries, although these sometimes have links with various forms of migration channels (Leiken 2004). Nevertheless, 'there is a growing tendency for governments to view long-term immigration objectives through the lens of security (Saggar 2008: 10).

There are significant consequences of the emphasis on securitisation because this is understood as 'an existential threat, requiring emergency measures and justifying actions outside the normal bounds of political procedure' (Buzan *et al.* 1998: 23–24). One manifestation of the securitisation of migration discourses is the notion of suspect communities, whereby particular migrant groups or ethnic minorities are presented as, or understood to be, sources of negative risks to particular societies. The positions of suspect communities have shifted over time and in the UK, for example, Muslims took over this unwanted mantle from the Irish after 2001. An analysis of representations of Muslims in the UK press demonstrates that, after this date, they became substantially more associated with terrorism and violence (Richardson 2004; Poole and Richardson 2006). Furthermore, Nickels *et al.* (2010) consider that British national identity is significantly based on the construction of 'others' in relation to race, ethnicity, religion and socioeconomic position. The communities which are presented as 'suspect' are usually those that are constructed as being most distant from the notion of Britishness.

Migration to Europe has increasingly been seen to constitute three types of threat (Carretero 2008). First, internal security threats as a result of the freedom of movement provisions of the Schengen Agreement, allowing migrants, asylum seekers and terrorists (the terms are often conflated) to move across internal borders without restriction. There also tends to be a conflation of crime and irregular migration in many of the discourses about the internal security of the EU (Koslowski 2001). Second, societal security in relation to cultural identities and welfare (Huysmans 2000; Adamson 2006), a notion which combines Ceyhan and Tsoukala's (2002) socioeconomic and identitarian axes. Thirdly, irregular migration is seen as a risk to external security (Adamson 2006; Sassen 2004; Sassen 1996). Irregular migrants become identified as potential terrorists in such discourses (Koser

2005), reflecting the loose, but far from accidental, conflation of terms. These highly ambivalent securitarian discourses are constructed around a series of myths (Ceyhan and Tsoukala 2002: 36), but they are increasingly dominant in debates about migration policies in developed countries. They are also often significantly detached from the realities of migration. Collyer (2010), for example, emphasises that irregular border crossings are relatively insignificant as a channel for entering Europe—and one that popular and policy discourses focus on—compared to the number of migrants who arrive with visas and overstay (de Haas 2008).

Although securitisation discourses are increasingly hegemonic, they are not monolithic. Economic discourses about the risks posed by migration also exist. Barysch (2005: 1), for example, reports the existence of widespread common perceptions in the older EU countries regarding 'unfair' labour market competition in the enlarged single market: 'Much of the resentment that has been building up in the old EU is fuelled by false perceptions about cheap Polish plumbers and Latvian builders "stealing" West European jobs by undercutting local wages and disregarding social standards'. But there are also counter-discourses, and many employers' organisations, for example, contend that to constrain labour migration is to increase the risks to economic growth, and that the spending of migrants' wages increases total demand and employment in the economy. For example, in 2012 John Cridland, the Chief of the UK business association, the Confederation of British Industry, criticised the UK government's aims to reduce net migration to the UK to 'tens of thousands' by 2015 because 'there's been so much rhetoric that it's creating its own reality, it's putting people off' (*Daily Telegraph*, 16 November 2012).

Another set of discourses revolves around ideas of modernisation. Flynn (2005) argues that, in the UK, New Labour sought to present an image of a radical new migration policy based on the notion of modernisation. It was linked to a political discourse about the importance of science and technology in the economy, privileging the interests of some sections of business, especially the internationally trading sectors, which value migration both as a source of knowledge and skills, and as a way of reducing costs to remain competitive (Finlayson 2003). There are also many competing discourses, including liberal and social-democratic discourses and imaginaries which emphasise the responsibilities of developed countries to less developed countries, both because of the interlinking of economic development, trade and migration, as well as humanitarian reasons. In practice, however, as will be seen in the next chapter, securitisation and other political dictates are also competing to shape migration policies that are best characterised as 'managed migration'. Moreover, although the securitarian and political axes tend to drown out the socioeconomic axis in popular and media discourses, they do not in any simple way dominate policy formation, not least because policy discourses in the public and private arenas, while linked, are not necessarily identical.

MIGRANTS AS VICTIMS

Although most discourses focus on the risks that are posed for societies by migration, there is a smaller but important set of discourses concerned with representations of migrants as victims. There are, of course, many migrants who have negative experiences of the migration process, as parts of their lives abroad, or in terms of the overall trajectory of their lives as a consequence of migration. These include the migrants who fail to find a job, are desperately homesick, are unable to bring their families to join them, or who become ill or die in the transit. There are some accounts which present them as victims, and this is often true of novels about their experiences, such as Steinbeck's (1939) *Grapes of Wrath*. Another example is Malachy McCourt, the husband and father in Frank McCourt's (1996) memoir, *Angela's Ashes*, who leaves Ireland to find work in England, and ends up jobless and sleeping rough. But these types of experiences are rarely the subjects of popular discourses on migrants as victims—perhaps because of the (often simplistic) assumptions made about the fact that such individuals 'chose' to migrate, and that the risks were presumably 'known'.

Instead, discourses about migrants as victims tend to focus selectively on 'illegal' migrants. Their representations tend to be polarised, as Anderson (2008: 2) explains.

> The 'illegal immigrant' is presented as either an exploited victim ('trafficked') or abuser of the system. This reflects a broader discourse on migration which separates foreigners into 'good' and 'bad' migrants: the hard working foreigner necessary for the economy, or the thief of jobs and opportunities.

Anderson stresses that these types of presentations are to be found not only in popular and media discourses, but also in policy discourses and in academic publications. Her review of UK immigration policy documents since 1998 demonstrates the pervasiveness of this dual representation of illegal migrants. They are presented as 'victims', mostly of traffickers, slavers and others, but occasionally as failures of the security forces or the welfare state to identify their conditions. Or they are presented as criminals, with irregular border crossing being conflated with criminality and potential terrorist threats. They also include the category of trafficker, that is, being the agents of exploitation of the trafficked.

Anderson (2008: 2) considers that the language of trafficking is significant, because it 'seems to help develop this into a more concrete opportunity to shift some "illegal immigrants" into the "good" category by recognising them as trafficking victims'. Politicians, journalists and editors are thus enabled to present themselves as simultaneously being against 'large-scale' immigration while also appearing to be humane and caring for the victims of trafficked migration. In this way, the 'solution' to 'the migration problem' is reduced into a series of actions to counter or eliminate traffickers—but effectively this removes both the 'bad' and the 'good' migrants from the picture. Given that prosecutions of traffickers or enslavers are rare, Anderson (2008: 4) considers that the main function of such discourses is in framing the debate about migration: 'In the case of trafficking and illegality the enforcement of immigration controls thus is presented as a means of protection of rights and in the interest of those who are "ruthlessly exploited"'.

One of the most emotive categories of victims of trafficking is probably migrant sex workers. A number of researchers, including Agustin (2007) have questioned the social categorisation of all migrant sex workers as unwilling victims, and she argues that this fails to recognise that many individual sex workers can and do take decisions about the risks associated with their work. It is not a job like any other job, but neither is it automatically an occupation where all the workers are passive victims. She argues that, on the one hand, there are those who emphasise the various negative risks faced by migrant sex workers, and deny agency to all the women involved. On the other hand, there are those 'who wish to recognize migrant women's agency attempt to normalize "sex work" to make this employment less risky and more socially acceptable but ignore the particular obstacles to agency posed by the illegality of most migrations' (Agustin 2005: 96). For Agustin, neither position is acceptable, reflecting the fact that the advocates in these debates are mostly social commentators from the more developed countries, and are not migrants. Those who seek to classify migrant sex workers as victims oversimplify their diverse experiences, and fail to capture the complexity of individual situations.

Agustin only partly engages with the notion of risk, but it is clear that this is significant in creating knowledge and 'truths' in relation to illegal migration, and in terms of social categorisation of victims. In a powerful exposition of how social categories are produced, Anderson (2008: 5) argues that 'the arguments become about state *identification* of victims, and the low numbers of those identified, as if they can be identified in the same way as blue-eyed people'. This is broadly in line with Foucauldian perspectives on how the creation of such categories is both informed, and given meaning to, by wider discourses in the media and elsewhere as to who are the 'real' victims of trafficking, thereby emphasising the distributed nature of power in society. Agustin (2005: 109) takes this argument further, contending that the categorisation of migrant sex workers as victims has political consequences. When women are presented as having been forced to come to, say, Europe to become sex workers, then a specific logic is created for justifying their repatriation. She reports that this logic has been used to repatriate—that is, to deport—individuals in France and Italy. The arguments about trafficking are also sometimes associated with the discourses about slavery (Box 9.3).

Box 9.3   Discourses about Slavery and Forced Labour

The ILO (2005) report, *A Global Alliance Against Forced Labour*, addresses the issue of contemporary slavery. Rogaly (2008) criticises the report for the way it conflates the notions of forced labour, slavery and slavery-type conditions. Similar criticisms have been noted elsewhere in this book relating to other aspects of the discourses about migrants, whether as victims or as risks to societies. Above all, Rogaly criticises the report for the way in which it deals with diversity and agency:

> The report explicitly subsumes large swathes of labour relations which contain elements of freedom as well as unfreedom, degrees of manoeuvrability, negotiation and contestation. . . . The language of 'victims' risks a descent into what Puwar has termed a 'politics of sympathy'.
> (p. 1444).

In contrast, he argues that many of the workers covered by this broad categorisation would not recognise their situation in the way that this implies.

For Rogaly, the discourse about victims has the effect of clouding what he considers to be the reality of the power relations that lie behind slavery and antislavery policies. Moreover, such a report does not simply appear as the outcome of 'objective' research, but is subject to intervention by the national governments which are represented on the board of the ILO. For example, the *Guardian* in 2005 reported that the UK board members had objected to the UK report on Forced Labour because it put the link between labour market deregulation and abuses under the spotlight.

Perhaps even more fundamentally, the discourse in the report about the role of 'private agents' in forced labour serves to distract attention from 'the role of contemporary forms of capitalism and their accommodation with states to produce both a lack of freedom and insecurity' (p. 1444). That is, it diverts attention from the political economy of employment conditions.

Source: Rogaly (2008)

CONCLUSIONS

Discourse is potent in shaping public and policy understandings, attitudes and practices in relation to migration, and seems to play a particularly powerful role in respect to irregular migration. The realities of migration—in terms of the numbers, destinations, economic roles, social inclusion, and the extent to which terrorism is associated with migration—are often distant from the ways in which these are portrayed in discourses. The state seeks to exert some control over the production of such discourses, but neither it nor the various forms of media have a monopoly of power in this respect. Instead, discourses are created and re-created, and are competing ways of seeing the world which are both influenced by and shape the realities of migration. Changes in the technology and the ownership or control of the media, and especially the growth of social media, have also led to changes in the distribution of control over discourses. Diverse popular discourses are often competing not only with each other, but also against a range of other discourses.

One of the most striking features of discourses is they way in which they can rapidly shift over time, as has been illustrated by the discourses about Japanese versus Chinese migrants in the USA in the mid-twentieth century. Similar shifts have occurred in the discourses about other migrant groups in both the USA and other countries. The shifting discourse about Irish migrants in the UK is a case in point. Other than nationality and ethnicity, the discourses are often centred on such cleavages as regular versus irregular, or economic versus political refugees. One of their most potent functions is the classification of types of migrants as deserving or undeserving, as being at risk or a risk. These discourses play a powerful role in the framing of migration policies, although not in any simple or deterministic manner, as is discussed in the following chapter.

10 Engaging with Risk in Migration Policies

INTRODUCTION

Discourses and policies are mutually framing in migration as in any other field. In terms of risk, migration is implicated in, and often central to, discourses and policies relating to issues such as health, crime, the environment and, above all, security. Migration poses risks to the ability of states to manage their territories in a number of ways, but especially the security of their borders in relation to what the migrants may be bringing with them, whether ideas, intentions or material. There is nothing new about this (Hollifield 2008), but the framing of policies has changed over time, in response to changing risks and discourses about risks, and in the nature of government and governance. In particular there has been an increasing focus on migration and risk in recent decades, and especially the nature of migration and associated risks. The national state has been the principal focus of migration policies, with a few exceptions such as the international refugee regime established by the Geneva Convention in 1950, and EU regulations in the area of asylum seeking and refugees.

According to Scott (1998: 88), states tend to have an 'aspiration to the administrative ordering of nature and society'. Over time this has generally led to increased state regulation to seek to impose this administrative ordering on areas as diverse as migration, policing, welfare and health. It is dependent on, and indeed is fed by the growth of, modern professional bureaucracies and new technologies that allow the monitoring and surveillance of particular threats or hazards (Carpenter 2001), and this has become central to migration policies, and their implementation. New technologies for monitoring migration have come into being in response to new policies, but—by modifying the capacities and powers of states—they also inform these. There is, however, a considerable gap between the framing of policies and their implementation. Best (2008: 360), for example, considers that there is a general tendency to be overoptimistic about the effectiveness of policies and strategies which seek to manage risks. In reality, understandings of the risks associated with migration, and indeed of the migration process itself, are characterised by gaps in knowledge and uncertainties about causes and consequences: Best (2008) writes that, even if we 'were to develop the most sophisticated of risk-management techniques, we would still be faced with the challenge of interpreting, debating, and communicating that information' (p. 361). risks are necessarily socially constructed while, in contrast, states tend to emphasise that it is 'calculable and apprehendable' (Best 2008: 363). In other words, there is a clash between the constructivist and objectivist perspectives.

The increasing focus on risk in migration is part of a much broader risk-aversion-informed policy shift. For example, regulatory policies in Europe have generally become more restrictive since the 1980s: 'They tend to be politicised, highly contentious and characterised by a suspicion of science and a mistrust of both government and industry' (Vogel 2001: 1). This shift is driven by a number of factors but two are particularly important. The first is the increasing public concern for health, safety and environmental regulation, as a result of greater exposure to popular and media discourse about the attendant risks. And the second is the increasing competence of the EU to regulate in these areas, which in turn has influenced the framing of national regulations. At the same time, media discourses around the failures of national and EU regulations, in areas such as the environment, have undermined public trust in government regulations, which in turn has led to increased regulation as states seek to reassert their legitimacy. As Hollifield (2008: 188) comments: 'If we accept the Weberian definition of sovereignty . . . a state can exist only if it has a monopoly of the legitimate use of force in a given territorial area. . . . It would then follow that the ability or inability of a state to control its borders and hence its population must be considered the sine qua non of sovereignty'.

These arguments are particularly relevant to migration, where there are strongly interlocking and mutually reinforcing relationships between state failures to secure borders, discourses about this, and a series of revisions (usually further tightening and selectivity) of migration rules. Policies and discourses tend to be segmented in relation to both the countries of origin, and the regulatory categories of the migrants: refugees and asylum seekers, and illegal migrants, are the dominant objects of attention.

THE STATE, RISK AND MIGRATION POLICIES

Two major themes have dominated migration policies over the very long term: economic and political expansion, and security risks. First, both ancient empires and premodern states used internal, and sometimes also international, population transfers to colonise sparsely populated areas and/or secure territorial gains: that is, to reduce the risk of their being lost to competing powers, and to reinforce their sovereignty. The other theme included concerns about invading nomads, risks of loosing scarce resources (land, food and water in particular), and internal security risks. The responses to these risks included tight border controls, and the construction of physical separation barriers. Examples included some of the most famous Roman defensive lines, such as Hadrian's Wall (see Box 10.1).

Box 10.1   Complex Relationships Between Risk and 'Migration Policies' in the Premodern Period

It was customary in the Roman Republic and Empire, from the fourth to the second centuries BC, to secure newly conquered territories by establishing colonies for veteran soldiers who would migrate to these places. Many of these colonies subsequently became prosperous cities that are still thriving today, such as York, Arles, and Köln. In Late Antiquity (fourth and fifth centuries AD), defence and security concerns dominated migration policies in the Roman Empire. The Roman Empire relied on military units of 'barbarian' tribes (*foederati*) to enforce their sovereignty in the face of external threats. In return, the *foederati* demanded the right to settle on Roman land. The Romans initially were hesitant about such land claims, but as the empire's tax revenue dwindled over time, the *foederati* were recognised as a low-cost alternative to maintaining a regular army (Goffart 2006: 238). Several Germanic tribes were allowed to settle in selected eastern and northern Roman provinces. The Emperor Valens, for example, allowed Goths to migrate to and settle in the province of Moesia (present day Bulgaria and Macedonia) in 376 AD. The Goths subsequently rebelled and defeated the Romans in 378 AD The disintegration of the Roman Empire followed three decades later.

In Imperial China, periods of active migration policies alternated with eras of isolationism. During the Sui (589–618) and Tang (618–907) dynasties, the imperial government encouraged the migration of ethnic minorities to settle in less populated areas of China (Wang 1995). However, states dominated by nomadic tribes conquered much of northern and central China during the Song dynasty (960–1279). The Mongolian Yuan Dynasty (1271–1368) was the first foreign dynasty to rule all of China, and treat the Han Chinese as second-class citizens. The Ming Dynasty (1368–1644) was a period of prosperity and population increases, so that there was no need for immigrants to populate China's territory. Instead, security concerns about continuing Mongol attacks dominated Chinese foreign policy. The Chinese government turned to isolationism. It abolished expensive naval expeditions and concentrated on building a strong army and extensive modernisation of the greatest of all physical barriers to population movement, the Great Wall of China.

Sources: Goffart (2006); Wang (1995)

However, there has been a marked shift in the regulation of borders in the last two hundred years. Drawing especially on Deleuze (1995), Urry (2007: 49) emphasises that states have long been concerned with monitoring mobile populations, especially those that cross its boundaries:

> State sovereignty is exercised upon territories, populations and, we may add, the movement of populations around that territory. The central notion is that of disciplining the 'ensemble of a population'. . . . From the early nineteenth century onwards governmentality involves not just a territory with fixed populations but mobile populations moving in, across and beyond 'territory'. The 'apparatuses of security' involve dealing with the 'population' but any such population is at a distance, on the move and needing to be statistically measured, plotted and trackable. . . . Such a 'mobile population' is immensely hard to monitor and to govern.

Although Hollifield (2008: 187) notes that the notion of tying populations to particular territories dates back to the sixteenth century, it was only in the nineteenth and twentieth centuries that systems of control and monitoring based on visas and passports were introduced, and significant measures were taken to close the border to those considered to represent a threat to the state. This was explicitly an attempt to mediate risks, a concept that was becoming increasingly evident in a number of policy domains in the nineteenth century (Taylor-Gooby and Zinn 2005; Ewald 1986). As migration policies developed during the course of the next two centuries, there was a strong convergence in migration policies in the more developed western societies (Myers 2002). While passports, visas, and various forms of legal measures were the main forms of restriction, physical barriers have also been important (Box 10.2). Drawing on Myers (2002) and Hollifield (2008), we can identify the following three key phases in modern migration policies in the developed countries.

Until the 1870s, immigration was largely unrestricted in most of the destination countries, which were significantly handicapped in terms of control by lacking the technologies for monitoring population flows and keeping these under surveillance. However, there were at least attempts to control immigration, with Britain, the United States, Canada, Switzerland and France all having introduced regulations by the 1790s. Over the next century, this was extended to most industrialised societies. In the last decade of the nineteenth century, up to the start of World War I, there was further tightening of controls in Australia, Britain, Canada, France, Germany, Sweden, and the USA, aimed principally at Eastern Europe and, in some cases, Japan.

Moving onto the post-World War II period, between 1945 and 1972 there was a phase of relatively open immigration policies, with Northern Europe attracting 'permanent' (or at least long-stay) immigrants from former

---

Box 10.2   Mobility, Risk and Physical Barriers

The oldest examples of border walls come from Rome and China, and the newest from the USA, Israel and India. Jones (2012: 3) argues that:

> Builders of barriers also sought to define who belongs within the state by creating and reifying boundaries both on the ground and in peoples' minds. In this way, the barriers are best understood as only the latest examples of sovereign states attempting to create a homogenised and orderly population inside a bounded territory.

Globally, there were some 30 physical separation barriers in the early 2010s, and most had been constructed in order to prevent illegal immigration.

Although the majority of border separation barriers were designed to prevent illegal immigration, a significant minority of such barriers have been constructed in order to prevent emigration. The most infamous barrier of this type was the former Iron Curtain separating European members of the former Warsaw pact from the NATO members, in the period 1948–1989. The Iron Curtain was a very efficient instrument for halting most East-West migration. According to Fassmann and Hintermann (1998: 60), there were only 13.3 million migrations westward between 1950 and 1990. Some 75% of these migration movements were under bilateral agreements for 'ethnic migration' and in most cases involved Jewish and German communities in the eastern bloc.

Sources: Jones (2012); Fassmann and Hintermann (1998)

---

colonies, Southern Europe, and refugees. States responded to the long post-war economic boom, and accepted—even encouraged—immigration as a means to increase the labour supply. Implicitly, the risks to the economy of failing to allow immigration were perceived to be greater than any associated negative risks that immigration involved. Despite this relaxation of border controls, and the attempt to incorporate migration into some of the international agreements to facilitate economic integration, migration flows across borders remained far more restricted than the parallel flows of goods and services (Adamson 2006: 165). It can be assumed that, to a large extent, this reflected the different risks perceived to be associated with the flows of goods as opposed to people across borders.

With economic recession, and more uncertain economic conditions from the early 1970s, the Northern European destination countries imposed much more restrictive immigration policies, as did Australia. The imposition of tighter immigration controls contributed to the growth of illegal or irregular migration, and thereafter this came to be seen as a major source of risk, and policy focus. A variety of measures were employed including stricter border controls, fines for employers of illegal immigrants, and harsher treatment of the migrants identified as being illegal. There were also reforms to the processing of asylum requests. There was not a complete ban on immigration, of course, because states have to balance competing interests, and contrasting risks, as discussed later in this chapter. But migration policies have become more restrictive, with priority being given to the highly skilled and to entrepreneurs, while family members also have relatively less restricted access. This has resulted not only in increases in illegal migration but also in asylum seeking, as the doors were closed to many potential migrants. There is, in short, a sharp polarisation of migrants according to their position in relation to migration regulation. The outcome, according to Collyer and de Haas (2012: 471), is that:

> Individuals who choose to travel with no documentation are separated from their documented counterparts at every stage of the journey and often for many years afterwards. They travel by different modes of transport on different routes; they must live in different places and they have different access to basic services; they take up different employment or the same employment for different rates of pay. It is of course possible to shift categorisations, through a variety of means, but in the context of migration to wealthy countries, the opportunities enjoyed by individual migrants are now significantly determined by their relationship with states.

It is not only categorisations and opportunities which have changed, but also the risks that migrants are seen to represent, and that they are exposed to, in these two very different migration worlds.

MIGRATION POLICIES IN THE LATE TWENTIETH CENTURY: THE TENSIONS BETWEEN A 'MOBILE WORLD' AND AN INTENSIFIED AWARENESS OF RISKS AND UNCERTAINTIES

The increased focus on the risks attendant on migration is reflected in the way that this has passed from being considered peripheral to the interests of international relations researchers, to becoming a major issue in the post-Cold War and post-9/11 period (Rudolph 2003: 31). There had, of course, been an earlier concern with migration risks, but these were understood differently. Whereas migration had previously been understood in terms of economic and social risks, in the early twenty-first century migration has come to be seen in terms of 'existential threats' (Walters 2008: 160). As Ceyhan and Tsoukala (2002: 23) note: 'By its transnational character, its dynamics, and its impact on people and institutions at all levels, migration is posing a serious challenge to the long-standing paradigms of certainty and order'. Governments have responded to these challenges in a number of ways, as outlined below, but they have also sought to balance security concerns against other goals, particularly the perceived economic advantages of a 'mobile world' (Walters 2004).

The War on Terrorism and New Forms of Risk Surveillance

In recent years, migration has assumed a hitherto unsurpassed importance in terms of the priority that has been accorded to it in security agendas in the developed world. There was growing concern in many developed countries about migration and the security of borders from the 1990s, and possibly from the end of the Cold War. For example, 50% of funds spent on technical assistance for the EU's *Phare* programme in Eastern Europe were spent on monitoring cross-border mobilities. However, the single event which was a catalyst to deepening this policy shift was the fact that a group of hijackers had entered the USA with the intentions of living and training there, and ultimately carrying out terrorist attacks on the World Trade Center and the Pentagon, on 11 September 2001. In subsequent years this pattern was apparently repeated in terrorist attacks in Europe, notably in Madrid and London, although in fact many of the perpetrators had been born in these countries, and were inaccurately portrayed as migrants.

In the USA, the government responded by significantly revising its security and border regulation practices. There had been strong forms of surveillance and monitoring in earlier time periods, notably the internment of Japanese populations during World War II (see Chapter Nine). However, in the early twenty-first century exceptional peace time controls were introduced, facilitated in part by technological developments in surveillance methods. In essence, there has been a sustained attempt in the USA—mirrored in many other developed western societies—to identify and draw lines around those considered to be a source of risk and those at risk (Edkins and Pin-Fat 2004; Levi and Wall 2004). As Amoore and de Goede (2008: 13–14) comment:

> Two worlds of globalization are represented through risk practice in the war on terror: one populated by legitimate and civilized groups whose normalized patterns of financial, leisure or business behaviour are to be secured; and another populated by illegitimate and uncivilized persons whose suspicious patterns of behaviour are to be targeted and apprehended. . . . In order that the licit and legitimate world of profitable movements of money, goods and people may remain an alluring and enduring prospect, then, control over the illicit world of terrorism, trafficking or illegal immigration must be made credible.

The EU, for example, deploys a digital database in order to monitor border crossings, building on earlier measures to use the Schengen Information System as a means of exchanging information amongst the Member States. However, the most publicised policy response was in the USA where the USA Patriot Act, passed in October 2001, introduced new powers of surveillance and detection of suspected terrorists, as well as expanding the definition of 'terrorist activities' in a way that facilitated the exclusion or detention of individuals. This was followed in May 2002 by the passing of the Enhanced Border Security and Visa Entry Reform Act, which significantly increased the number of immigration officers, and the scrutiny of entry visas. Male visitors from 'politically sensitive' countries, which were mostly Muslim, were also required to fulfill special registration procedures when entering the USA. Probably the most publicised aspect of the application of new technologies to identify migrants who are considered to constitute a risk was the adoption of biometric technologies to construct risk scores, under the 2004 US-VISIT programme and the 2006 Automated Targeting System (Achelpöhler and Niehaus 2004: 497). Other technological adaptations have included the introduction of the National Automated Immigration Lookout System, which matches individual details with databases of suspected international terrorists (Koslowski 2001).

Other national governments have adopted broadly similar practices; for example, the German government has screened data on more than five million individuals in an attempt to identify sleeper Al Qaeda cells. In the UK, a range of technologies has been utilised (Jennings 2007), including search and detection technologies such as carbon dioxide detectors and x-ray scanners to examine containers and other vehicles. Automatic number plate readers and closed circuit television are also widely used. At a European level, the Eurodac fingerprint database provides an international technology for border controls.

It is not just the technologies which have changed, but also how they are used. Faced with increasing risks and uncertainties, states have deployed new technologies to design risk practices that 'do not seek prevention, but pre-emption; they do not seek to reduce or limit risk, but to run with it; they are not designed to render safe or secure, but instead to give the appearance of securability'(Amoore 2008: 127). They are especially driven by the need to manage 'the unknown unknowns', or the uncertainties of terrorism (Aradau and van Munster 2008: 30). Managing uncertainty, of course, poses particular challenges, as you cannot rely on the logic of probabilities and calculable risks. Instead, more imaginative approaches have been utilised which rely on stress testing, scenarios and disaster rehearsal (O'Malley 2004: 5).

There has been relatively little resistance to the introduction of these new, more intrusive methods of surveillance, and to tighter border security. This is hardly surprising in context of the prevailing discourses about migration and terrorism in the wake of 9/11. A Fox News poll in November of that year indicated that 65% of Americans favoured preventing *any* immigration (Rudolph 2003: 616), and this was 'a clear indication of just how acute perceptions of threat had become and the clear linkage made between security and immigration control among the general public'. An especially close link was made between illegal migration and terrorism in many countries. For example, the Spanish Foreign Minister at that time, Josep Pique, pronounced that 'the fight against illegal immigration is also the reinforcement of the fight against terrorism' (quoted in Rudolph 2003: 616).

In this climate of heightened perceptions of risk and uncertainty, a number of governments introduced strong, sometimes draconian, measures to monitor, detain and exclude suspected terrorists, which ran counter to a previous, more liberal and, to some extent, broadly human rights informed approach to migration and border controls. In the United Kingdom, for example, 9/11 was followed by a significant extension of anti-terrorism measures (Flynn 2005): the Anti-Terrorism, Crime and Security Act of 2001 allowed the government to detain indefinitely suspected foreign nationals engaged in tourism. The 2002 Nationality, Immigration and Asylum Bill also increased the sanctions on employers and carriers for transporting those without appropriate paperwork, extended the prison sentences for trafficking, and increased powers to detain asylum seekers. There was also increased cooperation with the French government to try and reduce the use of the Channel Tunnel route by illegal migrants. But despite these security concerns, immigration policies continued to be informed as much by economic concerns as by security risks:

> The concept of a 'war against terrorism' had major implications for the type of managed migration schemes which the authorities were designing, and ideas about surveillance and intelligence-gathering activities among migrant movements would, in the months ahead, lead to a drive towards closer collaboration between UK, European Union, and United States police agencies. But the essential view that modern free market economies would be required to facilitate migration in some shape and form was preserved.
>
> (Flynn 2005: 475)

The state in the UK, as elsewhere in the developed world, continues to try and balance economic, social, and security risks and objectives. Nevertheless, the language of risk had firmly entered the policy arena in the UK (see Box 10.3), and there is no doubt that the balance in migration policies has shifted from economic and social objectives to increased security content. Many of these measures were introduced as temporary actions, but a decade later they already seem to have an air of permanence. However, there continues to be a significant gap between popular discourses and perceptions of the risks associated with migration, and migration policies, as discussed in the following section.

Box 10.3   Extract from the House of Commons Home Affairs Committee on Bulgarian and Romanian migration to the UK, 2008

On 27 November 2007 Liam Byrne, the UK's Minister of State for Borders and Immigration, addressed the House of Commons select committee on migration from the—then—newest member states of the EU, Bulgaria and

Romania. His language was infused with the notions of risk, in this case economic and social risks:

> I do not think that we should take any *risks* in migration policy at the moment. If we lifted those restrictions overnight in a blanket way we would create new *risks*. When I had the Migration Impact Forum report to me that there were specific isolated concerns about pressure on public services my advice to the Home Secretary was that we should not be taking any *risks*. I now have the figures for both child benefit and tax credits. There were 634 child benefit claims of which 380 were approved. There were 195 tax credit claims of which 137 were awarded.
>
> (emphases added)

Responding to further questions about allegations that migrants were claiming benefits in the UK, but then returned to live in their own countries, his response again utilised the language of risk:

> I do not believe I have been told that formally but, like you, that is a *risk* in the system of which I am aware. It is for DWP [Department of Work and Pensions] to police it and that is the kind of issue we talk about when we weigh up whether or not restrictions should continue.

Source: House of Commons Home Affairs Committee (2008)

MIGRATION AND RISK: THE DIVERGENCE BETWEEN POPULAR DISCOURSES AND POLICIES

In most western developed economies there is a substantial gulf between migration policies and popular discourses, or mass public opinion about migration (Massey 1998). For example, Lahav and Guiraudon (2006: 203) write that 'there has been a discrepancy between the desires of a largely anti-immigration public and the expansive bias of policies'. It is a gap which reflects the influence of particular interest groups, especially of capital as opposed to the general public. These reflect their different engagements with the ways in which the '"national interest" of states can be defined largely along three dimensions: (1) geopolitical security, (2) the production and accumulation of material wealth, and (3) social stability and cohesion' (Rudolph 2003: 604). Moreover, states construct 'grand strategies' that cut across all three of these domains.

It is in some ways obvious why the construction of the grand strategy for migration in the developed world has changed post-9/11. More intriguing, and a puzzle, for Rudolph is the construction of the grand strategy prior to this event. In the two preceding decades there had been an encroaching neo-liberalism in the grand economic strategies of many, perhaps most, western developed economies, but these applied to trade and capital mobility rather than labour migration. In short:

> Openness toward migration flows makes perfect sense given these links among migration, the labor demands of the new global economy, trade, and the pursuit of economic maximization. Surprisingly, however, while the contemporary logic of trade and international finance has been one of generally increasing degrees of openness, the logic of migration policies has increasingly been one of closure since the mid-1960s.
>
> (Rudolph 2003: 604)

In contrast, after 9/11 there was an increased awareness of migration as a channel of geopolitical security risks. While much of the discourse around this (Chapter Nine) linked the risks especially to illegal migration, in reality most of the major terrorist events in the twentieth century which involved 'migrants' involved those who were regular migrants. There have been attempts to increase migration controls since 9/11, as noted earlier in this chapter, including the application of new surveillance technologies, but there has not been a major reduction in migration and visitor flows across most international borders. This therefore poses a second puzzle: why in these circumstances are states willing to '"risk migration" and accept "unwanted immigrants"'? (Hollifield 2008: 84).

The answer to this is, of course, complex, and includes the difficulties states face in negating existing rights (for example, to family reunification migration), and the ways in which networks effectively mediate the increased costs of migration resulting from tighter border regulation. Judiciaries have intervened in a number of countries, including the UK, to block policies which are considered illegal or unconstitutional. As Hollifield (2008: 211) comments: 'Rights have a very long half-life in liberal democracies. Once they are extended and institutionalized, it is extremely difficult to roll them back'. However, arraigned against these are a number of powerful myths about national identities, around which are constructed 'symbolic politics' (King 2005) of the 'need to control migration'. However, states do resist pressures to restrict migration, in part because of the interest group pressures exerted by (some) employers. That brings us back to the question of why the symbolic politics of 'uncontrolled migration' have not resulted in much stronger, and even more restrictionist, immigration policies.

The answer to the above questions, of course, lies in the power of the business interests associated with 'the production and accumulation of material wealth'. Their voices, and sympathetic elements within the state, were able to keep the migration doors relatively open in the postwar decades even though there was a growing popular discourse about the need to control migration. And those same interests have been able to insist that the increased control and surveillance of migration post-9/11 was not articulated in the form of much more restrictionist policies.

In the UK, for example, Flynn (2005: 465) considers that 'the elements that prevail throughout New Labour's [in the 2000s] version of managed migration shows a system which is dominated by business interests'. More often, however, states tend to resort to 'intentional incoherence of policies' (Boswell 2007), which fogs the policy-making sphere. In Germany this takes the form of exceptionalism. In 1973 the country declared there would be an effective end to the recruitment of labour migrants but 'unpublicized and complex administrative arrangements have allowed for multiple exceptions' (Boswell 2007: 95). There are, however, limits to the strategy of exceptionalism, because ultimately the gap between the supposed de jure and the actual de facto policies can lead to a failure of legitimisation. Italy's government also adopted a strongly restrictionist rhetoric in the 2000s, while in practice it tolerated high levels of irregular migration, which were post-legitimated via periodic regularisation programmes. Such programmes were often implemented in the name of risk reduction—reducing risks to irregular migrants stemming from their legal status.

REFUGEES AND RISK

Globally, there were an estimated 12 million refugees in 2001, almost one half of whom were in Africa. Refugees are often simplistically categorised into two types—economic and political—in both popular and policy discourses. Indeed, Adamson (2006: 173) considers that:

> International law distinguishes between political and economic migration by assigning categories to individuals who are seeking to cross borders to escape political persecution or violent conflict, as opposed to those who cross borders in search of economic opportunities. International law defines 'refugees' as those who have a well-founded fear of persecution because of race, religion, nationality, or membership in a particular social or political group.

Although Adamson does not use the language of risk, in many ways forced migrations epitomise the relationship between migration and negative risks.

There is a long history of forced migration related to conquest and devastation wreaked by invading forces. Many of the major migrations throughout history have occurred as a result of some form of disaster or expulsion. One of the earliest was the creation of a Jewish diaspora after the destruction of the Temple of Jerusalem in the sixth century BC, but there have been numerous refugee movements driven by famine or persecution (Adamson 2006). Slavery was another source of enforced migration and it is estimated that some 15 million Africans were enslaved and forcibly transferred to the Americas before circa 1850. Jews were also the subject of persecution in, for example, late nineteenth century Russia and Hitler's Germany, leading to large-scale flights. More recently, the term 'ethnic cleansing' was coined to describe the persecutions that led to large-scale migrations in the former Yugoslavia in the 1990s. And away from Europe, 'ethnic cleansing' has been the driver of enormous refugee flows in Rwanda and other countries. The persecuted groups in these refugee flows are variously presented in discourses as being both at risk, and a risk to some other body whether it be the state or the ethnic majority (and sometimes minority) population.

Most refugees travel relatively short distances, perhaps because they lack the resources for long-distance movements, perhaps because they believe they will, or wish to, return as soon as possible, or because they do not want to run the risk of disrupting valuable support networks. Not surprisingly, therefore, although most of the focus on refugees has been on relatively high income destination countries, most refugees are found in relatively poor neighbouring countries, especially in Africa. Another important concentration of refugees includes the Palestinian refugee camps in Lebanon.

More recently, environmental refugees—mostly from climate change—have attracted attention as being the focus of various forms of risk, as noted in Chapter Eight. The fifth Global Forum on Migration and Development (GFMD) in Mexico in 2008 addressed some of these risks, although the conclusions did not advance very far beyond calling for more dialogue (Black *et al.* 2012). Migration is one response in the face of environmental threats, but as such migrations engender enormous risks for the migrants and their destinations, many of those at risk decide not to move and to seek ways to adapt. There have been similar discussions about how to design policies that will facilitate staying rather than moving. However, if forced migration is impossible to avoid, there is recognition of the need to create a framework for climate change resettlement which reduces the risk to all parties (de Sherbinin *et al.* 2011: 457). Black *et al.* (2012), drawing on Lebel *et al.* (2010), illustrate such adaptation with the example of Monsoonal Asia. Here migration is one possible means of adaptation, alongside strategies of immobility which includes constructing houses on stilts in order to coexist with the risks from floods. This is considered to be more viable than major engineering responses, such as dam building, which encourage people to live in places of very high risk, should the design capacity of the dams be exceeded. Drawing on Etkin (1999), Black *et al.* (2012) consider this to be an example of 'risk transference' to infrequent but high-impact events from frequent but low-impact events.

These risks are recognised as being significant. For example, the Inter Agency Standing Committee (IASC) Guidelines on the Protection of Persons in Situations of Natural Disasters has identified that displacement potentially puts significant numbers at risk (IASC 2011: 2). However, there remain considerable challenges in designing effective policies to engage with such risks, let alone to implement such policies effectively. These include lack of knowledge about risks, lack of shared understandings of risk, and lack of economic resources or political means to implement particular strategies.

Box 10.4   Asylum Seekers, Risk and Uncertainty in the UK

Jennings (2007) writes that although asylum has long been recognised in international law, there has been a change in the policy language relating to asylum in recent years, with the infusion of the terminology of risk and risk management.

A refugee is defined in the 1951 UN Convention as someone who, 'owing to well-founded fear of being persecuted for reasons of race, religion, nationality, membership of a particular social group or political opinion, is outside the country of his nationality and is unable or, owing to such fear, unwilling to avail himself of the protection of that country' (UN Convention 1951, Article 1A). The notion of well-founded resonates with the notion of certainty, and it is only more recently that uncertainty has entered policy implementation.

Faced with growing numbers of asylum seekers, in 1996 the UK parliament gave power to the Secretary of State to identify countries where there was 'in general no serious risk of persecution' (Section 1, c. 49). The 1996 Act also denied the right of appeal to asylum applicants who had travelled to the UK via what were considered to be 'safe' third countries. These powers became even more restrictive in 2002 when further legislation removed the right of appeal from applicants from particular countries whose cases were considered to be 'clearly unfounded' and 'not to be at serious risk of persecution' by the Secretary of State. Greater use has also been made of transit visa regimes in order to both deflect, and deter, asylum claims by 'high risk' visitors to the UK.

Furthermore, as in other areas of migration regulation, there has been greater reliance on technological solutions in responses to what are perceived to be risks associated with the process of asylum seeking. The UK government White Paper, *Fairer, Faster and Firmer* (Home Office 1998), concentrated resources on what were considered to be the areas of greatest risk. The technologies of control include detection, identity verification and information management. These rely on risk assessment, combined with the imposition of restrictions on specific national and other groups.

Source: Jennings (2007)

Asylum seekers are also associated with particular risks that have attracted the attention of migration policy makers, as illustrated by the case of the UK (Box 10.4).

CONCLUSIONS: REFLECTIONS ON MIGRATION POLICIES AND RISK

Migration policies are shaped by the interaction between competing interests, and much of this competition is played out in the form of policy and media discourses, as discussed in Chapter Nine. Those interests include the powerful voices of employers, for whom migrants represent an important source of labour, and potentially a means to enhance their competitiveness. The competing interests include those centred on housing, public services, cultural and national identities. These are strongly class based, and often territorially based—for example, in the UK the interests of London often seem to diverge from those of much of the remainder of the UK. The state is not a neutral vehicle in what is sometimes a veritable storm of competing interests, but rather it has agency in respect to migration and risk. It is inhabited by political parties, which do not in any simple way reflect or mediate popular discourses, and by bureaucracies with their own interests and values. The state therefore does not just respond to popular and policy discourses, but actively shapes these through pronouncements and its regulatory activities.

During most of the post-World War II period, the 'grand strategy' of migration policies in the developed western economies was largely informed by a neoliberal, international integration perspective (Rudolph 2003: 618). However, as we have noted, migration policies tended to be more constraining than other internationalisation policies, such as those for trade and investment. In other words, while risks to economic interests were dominant, they were challenged by social and cultural interests, many of which centred on discourses about risks to national identities. Subsequently, the rise of various forms of counter-globalism, including discourses about risks to employment, housing and welfare systems, have constrained the ability of the state to prioritise economic interests (Rosenau 1997). The outcome was a reconfiguration of the grand strategy, and this became even more marked with the increased awareness of security risks after 9/11.

> Policy developments during the 1990s represented government attempts to address societal insecurities through highly symbolic policies that present a strong *image* of control; however, constraints on neoliberal policies have become even more significant and increasingly difficult for states to 'finesse' as military security interests have converged with societal interests after 9/11. Indeed, the potential costs of the migration-terrorism link call into question whether it is in the national interest to attempt to craft policy toward assuaging *fears* rather than establishing stringent control over *flows*.
>
> (Rudolph 2003: 618; emphasis in the original)

In practice, most states have combined control measures with attempts to assuage popular fears about the risks of migration. The latter takes the form of the contributions of politicians to policy and popular discourses, and the passing of measures that are meant to reduce the risks associated with migration. Often this is more about visibility and gestures rather than outcomes. The latter are far more problematic, as we discuss below. However, this is not to deny that states do exercise significant controls over borders and surveillance of 'suspect' migrant populations. On the contrary, in response to increased security risks, which were strongly associated with migration, even when the risk in some cases came from non-migrants, the western developed countries have significantly reinforced border controls in recent decades, especially since 9/11. This has been facilitated by technological developments, but, at the same time, traditional material barriers—walls and fences—are still being constructed as ways of minimising perceived risks to the destination countries. For example, the budget of the US immigration control service tripled in the decade after 2003, physical barriers have been built or reinforced at the border with Mexico, and the number of agents employed in border controls doubled (Adamson 2006: 178). Surveillance remains labour intensive, although the outstripping of employment growth by total expenditure is indicative of the scale of investment in new technologies.

The tightening of border controls and surveillance carries significant costs. Firstly, there are the direct financial costs of the increased controls and surveillance, which Rudolph (2003: 618) contends can be prohibitive for states with long land borders. Secondly, there are the costs for the economies resulting from the loss of the positive contribution of migration to labour supply conditions for firms. Thirdly, there are costs 'in civil liberties in order for governments to separate friend from foe' (Rudolph 2003: 619). Fourthly, there are the costs borne by the migrants themselves, who often have to face even greater risks as they turn in desperation to ever more dangerous ways of crossing borders. This is evidenced by the experiences of African migrants trying to enter Europe via highly dangerous boat journeys in the face of increased surveillance by the Spanish, and other, states (Carling 2007). As Collyer (2010: 5) writes, the most significant effect is probably that 'the displacement impact of greater border controls at easily accessible border points obliges migrants to travel across more hostile terrain'. Most migrants have some understanding of the risks attached to border crossings, but—for a variety of reasons—decide to tolerate those risks. Death or serious injury is, of course, not the only risk encountered in such border crossings. There is also the risk of being cheated by the intermediaries who promise to arrange border crossings, especially as this is often a domain where levels of trust are low, and difficult to enhance, once the migrants have left their home countries. Another risk is that of becoming a stranded migrant (Collyer 2010). Irregular migrants travelling by land and sea, rather than by air, are likely to have journeys characterised by fragmented risks, as they traverse different national regulatory regimes. They may traverse some regimes successfully, but then become stranded or entrapped when they cannot evade the controls in one of the intermediary countries on the routes to their destinations.

Whatever the risk-managing intentions of migration policies, the outcomes—even in terms of these objectives—tend to be complex and uneven. There are several reasons for this, including the way in which accumulated migration regulations have, over the years, tended to produce complex categorisations of migrants. Koffman (2002) provides an overview of the rights of different categories of migrants in the EU, and Morris (2002) identified 25 categories of legal residence status for migrants in the UK, which underlines the oversimplification involved in any simple bipolar categorisation of regular versus irregular migrants. These categorisations are not explicitly defined in terms of risk, but risk is implicit in the decisions regarding the rights and exclusions which they define. These categorisations also define the risks faced by individual migrants.

Another reason for the uncertain outcomes of regulations is the limits to the powers of states to control their borders. Despite attempts by states to secure sole control over border crossings, via passport and visa regimes (Torpey 1999), in practice they are unable to maximise, let alone monopolise, control over such movements. This is particularly true in western liberal democracies, where there are counteracting human rights pressures, but is also true of authoritarian states. Even North Korea cannot completely control cross-border movements. There are a number of reasons for this including the sheer impossibility of detecting every possible cross-border crossing, or at least the intentions of those who cross these borders (sometimes as regular migrants, with short-term visas). There is also a small army of intermediaries, traffickers and smugglers, who have competing powers to those of the state and can mediate the risks involved in border crossings (see Chapter Seven). Adamson (2006: 179) noted that one of the perverse effects of increased border controls can be to reduce the financial costs of using intermediaries, in some circumstances. For example, he reports that the fees paid to smugglers who assisted migrants to move from Mexico to Arizona fell by about one half after 9/11 because there was an increased risk (probability) of being detected. He quotes one border crosser as saying that: 'Because of this bearded guy, what's his name, bin Laden, it is harder now. There are more reinforcements now because America is afraid of terrorism'. Another reason for the limits to the ability of states to control their borders is the high level of willingness to take risk amongst migrants, whether because this reflects their desperation, or because it is linked to their perceived competence to manage those risks—whether individually, or through the resources of smugglers. In fact, border management agencies rarely have any detailed understanding of the willingness of migrants to take risks, which is a major lacuna. Nieuwenhuys and Pécoud (2007) consider that border agencies often assume—wrongly—that irregular migrants are unaware of the risks they are taking, but it is just as likely, or even more likely, that they are aware of these risks, but willing to tolerate them, or believe they can be managed.

Finally, there is one other important aspect of willingness to take risks when crossing borders, and this is the distinction between risk and uncertainty (Knight 1921). Do migrants understand border crossings as involving risk (known risks) or uncertainties (unknown risks)? It may well be the case that uncertainty tolerance is more important than risk tolerance in this respect. Similarly, migration policy makers are trying to regulate future mobilities, the determinants of which are partly known and constitute risks, but they also face enormous uncertainties, relating especially to conditions in potential countries of origin.

Bibliography

Abbot, D., Jones A. and Quilgars, D. (2006), 'Social inequality and risk', in Taylor-Gooby, P. and Zinn, J. O. (eds.), *Risk in social science*, Oxford: Oxford University Press, pp. 228–249

Abdellaoui, M. (2000), 'Parameter-free elicitation of utility and probability weighting functions', *Management Science*, 46: 1497–1512

Abdellaoui, M., L'Haridon, O. and Bleichrodt, H. (2008), 'A tractable method to measure utility and loss aversion under prospect theory', *Journal of Risk and Uncertainty*, 36: 245–266

Abdellaoui, M., L'Haridon, O. and Paraschiv, C. (2011), 'Experienced vs. described uncertainty: Do we need two prospect theory specifications? *Management Science*, 57: 1879–1895

Abdellaoui, M., Vossmann, F. and Weber, M. (2003), *Choice-based elicitation and decomposition of decision weights for gains and losses under uncertainty*, CEPR, Discussion Paper No. 3756

Achelpöhler, W. and Niehaus, H. (2004), 'Data screening as a means of preventing Islamist terrorist attacks in Germany', *German Law Journal*, 5: 495–513

Adam, B. and van Loon J. (2000), 'Introduction', in Adam, B., Beck, U. and van Loon, J. (eds.), *The risk society and beyond: Critical issues for social theory*, London: Sage, pp. 1–31

Adams, J. (1995), *Risk*, London: Routledge

Adams, J. (2001), *Risk*, London: Routledge

Adamson, F. B. (2006), 'Crossing borders: International migration and national security', *International Security*, 31: 165–199

Agardi, T., Alder, J., Dayton, P., Curran, S., Kitchingman, A., Wilson, M., Catenazzi, A., Restrepo, J., Birkeland, C., Blaber, S., Saifullah, S., Branch, G., Boersma, D., Nixon, S., Dugan, P., Davidson, N. and Vorosmarty, C. (2005), 'Coastal systems', in Hassan, R. M., Scholes, R. and Ash, N. (eds.), *Millennium ecosystem assessment: Ecosystems and human well-being: Current State and Trends*, vol. 1, Washington, D.C.: Island Press, pp. 513–549

Agustin, L. (2005), 'Migrants in the mistress's house: Other voices in the trafficking debate', *Social Politics: International Studies in Gender, State and Society*, 12: 96–117

Agustin, L. (2007), *Sex at the margins: Migration, labour markets and the rescue industry*, London: Zed Books

Akerlof, G. A. (1970), 'The market for "lemons": Quality uncertainty and the market mechanism', *The Quarterly Journal of Economics*, 84: 488–500

Alexander, J. (1996), 'Critical reflections on "reflexive modernisation"', Theory, Culture and Society, 13: 133–138

Alheit, P. (1994), 'Everyday time and life time', *Time and Society*, 3: 305–319

Allen, J. M. and Eaton, B. C. (2005), 'Incomplete information and migration: The grass is greener across the higher fence', *Journal of Regional Science*, 45: 1–19

Alwang, J., Siegel, P. B. and Jorgensen, S. L. (2001), *Vulnerability: A view from different disciplines*, New York: The World Bank, Social Protection Discussion Paper Series No. 0115

Amin, A. (2002), 'Spatialities of globalization', *Environment and Planning A*, 34: 385–399

Amoore, L. (2008), 'Consulting, culture, the camp: On the economies of the exception', in Amoore, L. and de Goede, M. (eds.), *Risk and the war on terror*, London: Routledge, pp. 112–129

Amoore, L. and de Goede, M. (2008), 'Introduction: Governing by risk in the war on terror', in Amoore, L. and de Goede, M. (eds.), *Risk and the war on terror*, London: Routledge, pp. 5–19

Anderson, B. (2000), *Doing the dirty work: The global politics of domestic labour*, London: Zed Books

Anderson, B. (2008), *"Illegal immigrant": Victim or villain?*, Oxford: University of Oxford, COMPAS Working Paper 64. Retrieved from: www.compas.ox.ac.uk/publications/working-papers/wp-08-64/.n [Accessed October 2012]

Anheier, H. and Kendall, J. (2002), 'Interpersonal trust and voluntary associations: Examining three approaches', *The British Journal of Sociology*, 53: 343–362

Annis, D. B. (1987), 'The meaning, value and duties of friendship', *American Philosophical Quarterly*, 24: 349–356

Aoki, K. (1996), 'Foreign-ness & Asian American identities: Yellowface, World War II propaganda and bifurcated racial stereotypes', *UCLA Asian Pacific American Law Journal*, 4: 1–71

Aradau, C. and van Munster, R. (2008), 'Taming the future: The dispositive of risk in the war on terror', in Amoore, L. and de Goede, M. (eds.), *Risk and the war on terror*, London: Routledge, pp. 23–40

Arrow, K. J. (1964), 'The role of securities in the optimal allocation of risk-bearing', *Review of Economic Studies*, 31: 91–96

Australia Government Department of Immigration and Citizenships (2013). Available at: www.immi.gov.au/visas/migration-agents/unregistered-agents.htm [Accessed October 2013]

Aven, T. (2007), 'A unified framework for risk and vulnerability analysis and management covering both safety and security', *Reliability Engineering and System Safety*, 92: 745–754

Aven, T. and Renn, O. (2009), 'On risk defined as an event where the outcome is uncertain', *Journal of Risk Research*, 12: 1–11

Baines, D. (1995), *Emigration from Europe 1815–1930*, Cambridge: University of Cambridge Press

Baines, D. (1998), 'European emigration, 1815–1930: Looking at the emigration decision again', *The Economic History Review*, 47: 525–544

Baláž, V., Bačová, V., Drobná, E., Dudeková, K. and Adamík, K. (2013), 'Testing project theory parameters', *Ekonomický časopis*, 61: 655–671

Baláž, V., Fifeková, E. and Nemcová, E. (2009), 'Ellsberg paradox: Decision-making under risk and uncertainty', *Ekonomický časopis*, 57: 213–229

Baláž, V. and Williams, A. M. (2004), '"Been there, done that": International student migration and human capital transfers from the UK to Slovakia', *Population, Space and Place*, 10: 217–237

Baláž, V. and Williams, A. M. (2011), 'Risk attitudes and migration experience', *Journal of Risk Research*, 14: 583–596

Baláž V. Williams A. M. and Fifeková, E. (2014), 'Migration and complex choice. Looking into migrants' minds', forthcoming in: *Population, Place and Space* doi: 10.1002/psp.1858

Baláž, V., Williams, A. M. and Kollár, D. (2004), 'Temporary versus permanent youth brain drain: Economic implications', *International Migration*, 42: 3–34

Banerjee, A. V. (1992), 'A simple model of herd behaviour', *Quarterly Journal of Economics*, 107: 797–817

Barberis, N., Huang, M. and Santos, T. (2001), 'Prospect theory and asset prices', *Quarterly Journal of Economics*, 116: 1–53

Barsky, R., Juster, T., Kimball, M. and Shapiro, M. (1997), 'Preference parameters and behavioral heterogeneity: An experimental approach in the health and retirement study', *Quarterly Journal of Economics*, 112: 537–579

Barth, F. (1969), 'Introduction', in Barth, F. (ed.) (1969), *Ethnic groups and boundaries: The social organization of cultural differences*, London: George Allen and Unwin, 9–38

Barysch, K. (2005), *East versus west? The European economic and social model after enlargement*, London: Centre for European Reform Essays

Bauer, T. (1995), *The migration decisions with uncertain costs*, Munich: Ludwig-Maximilians-Universität München, Münchener Wirtschaftswissenschaftliche Beiträge, No. 95-25

Bauer, T., Epstein, G. S. and Gang, I. N. (2007), 'The influence of stocks and flows on migrants' location choices', *Research in Labor Economics*, 26: 199–229

Bauman, Z. (1987), *Legislators and interpreters: On modernity, postmodernity and intellectuals*, Ithaca, NY: Cornell University Press

Bauman, Z. (2001), The individualized society, Cambridge: Polity Press

Bayes, T. and Price, R. (1763), 'An essay towards solving a problem in the doctrine of chances', *Philosophical Transactions*, 53: 370–418

Beck, U. (1992), *Risk society: Towards a new modernity*, London: Sage

Beck, U. and Beck-Gernsheim, E. (2002), *Individualization*, London: Sage

Becker, G. A. (1991), *Treatise on the family*, Boston: Harvard University Press

Beine, M., Docquier, F. and Rapopor, H. (2001), 'Brain drain and economic growth: Theory and evidence', *Journal of Development Economics*, 64: 275–289

Bellaby, P. (1990), 'To risk or not to risk? Uses and limitations of Mary Douglas on risk-acceptability for understanding health and safety at work and road accidents', *Sociological Review*, 38: 465–483

Benjamin, D. and Brandt, L. (1998), *Administrative land allocation, nascent labor markets, and farm efficiency in rural China*, Toronto: University of Toronto, Department of Economics Working Paper (quoted in Taylor and López-Feldman 2007)

Benson-Rea, M. and Rawlinson, S. (2003), 'Highly skilled and business migrants: Information processes and settlement outcomes', *International Migration*, 41: 59–79

Bentham, J. (1789), *An introduction to the principles of morals and legislation*, first published in London: T. Payne and Son, reprinted by Dover Publications 2007

Berkeley, R., Khan, O. and Ambikaipaker, M. (2006), *What's new about the new immigrants in twenty-first century Britain?* Joseph Rowntree Foundation, York: York Publishing Services, Ltd.

Bernoulli, D. (1986), 'Exposition of a New Theory on the Measurement of Risk', *Econometrica*, 54(1): 23–36, translated from Latin original work *Specimen theoriae de mensura sortis, Commentarii Academniae Scientiarum Imperialis Petropolitanae*, (1738), Vol II, 175–192

Best, J. (2008), 'Ambiguity, uncertainty, and risk: Rethinking indeterminacy', *International Political Sociology*, 2: 355–374

Bettman, J. R., Luce, F. M. and Payne, J. W. (1998), 'Constructive consumer choice', *Journal of Consumer Research*, 3: 187–217

Betts, S. C. and Taran, Z. (2006), 'A test of prospect theory in the used car market: The non-linear effects of age and reliability on price', *Academy of Marketing Studies Journal*, 10: 57–75

Bhugra, D. (2004), 'Migration and mental health', *Acta Psychiatrica Scandinavica*, 109: 243–258

Bikhchandani, S., Hirshleifer, D. and Welch, I. (1992), 'A theory of fads, fashion, custom, and culture change as informational cascade', *Journal of Political Economy*, 100: 992–1026

Bilger, V., Hofmann, M. and Jandl, M. (2006), 'Human smuggling as a transnational service industry: Evidence from Austria', *International Migration*, 44: 60–93

Birnbaum, M. H. and LaCroix, A. R. (2008), 'New paradoxes of risky decision making', *Psychological Review*, 115: 463–501

Black, R., Adger, N., Arnell, N. W., Dercon, S., Geddes, A. and Thomas, D. (2011a), 'The effect of environmental change on human migration', *Global Environmental Change*, 21S: S3–S11

Black, R., Arnell, N. W., Adger, N., Thomas, D. and Geddes, A. (2012), 'Migration, immobility and displacement outcomes following extreme events', *Environmental Science & Policy*, 27S: S32–S43

Black, R., Kniverton, D. and Schmidt-Verkerk, K. (2011b), 'Migration and climate change: Towards an integrated assessment of sensitivity', *Environment and Planning A*, 43: 431–450

Blackler, F. H. M. (2002), 'Knowledge, knowledge work and organizations', in Choo, C. W. and Bontis, N. (eds.), *The strategic management of intellectual capital and organizational knowledge*, New York: Oxford University Press, pp. 47–64

Bleichrodt, H. and Pinto, J. (2000), 'A parameter-free elicitation of the probability weighting function in medical decision analysis', *Management Science*, 46: 1485–1496

Blood, R. O., Jr. (1962), *Marriage*, New York: Free Press

Boholm, A. (2003a), 'The cultural nature of risk: Can there be an anthropology of uncertainty?', *Ethnos*, 68: 159–178

Boholm, A. (2003b), 'Situated risk: An introduction', *Ethnos*, 68: 157–158

Böcker, A. (1994), 'Chain migration over legally closed borders: Settled immigrants as bridgeheads and gatekeepers', *The Netherlands Journal of Social Science*, 30: 87–106

Böcker, A. (2008) 'International retirement migration: The legal framework and its effects on migrants' choices and behaviour', in Balkir, C. (ed.), *Uluslararası Emekli Göçünün Ekonomik ve Sosyal Etkileri:Antalya Örneği*, Antalya: Büyükşehir Belediyesinin katkılarıyla basılmıştır, pp. 105–109.

Borjas, G. J. (1990), *Friends or strangers: The impacts of immigrants on the US economy*, New York: Basic Books

Boswell, C. (2007), 'Theorizing migration policy: Is there a third way?', *International Migration Review*, 41: 75–100

Boucher, G. (2008), 'A critique of global policy discourses on managing international migration', *Third World Quarterly*, 29: 1461–1471

Bourdieu, P. (1984), *Distinction*, London: Routledge and Kegan Paul

Boustan, L. P., Kahn, M. E. and Rhode, P. W. (2012), 'Moving to higher ground: Migration response to national disasters in the early twentieth century', *American Economic Review*, 102: 238–244

Boyd, M. (1989), 'Family and personal networks in international migration: Recent developments and new agendas', *International Migration Review*, 23: 638–670

Boyd-Bowman, P. (1976), 'Patterns of Spanish emigration to the Indies until 1600', *Hispanic American Historical Review*, 56: 580–604

Brettel, C. B. (2008), 'Theorizing migration in anthropology: The social construction of networks, identities, communities and globalscapes', in Brettell, C. B. and Hollifield, J. (eds.), *Migration theory: Talking across disciplines*, New York: Routledge, pp. 113–159

Brettell, C. B. and Hollifield, J. (2008), 'Migration theory: Talking across disciplines', in Brettell, C. B. and Hollifield, J. (eds.), *Migration theory: Talking across disciplines*, New York: Routledge, pp. 1–30

Bruegel, I. (1996), 'The trailing wife: A declining breed?', in Crompton, R., Gaillie, D. and Purcell, K. (eds.), *Changing forms of employment*, London: Routledge, pp. 235–259

Brunnermeier M. K, and Nagel, S. (2005), *Do wealth fluctuations generate time-varying risk aversion? Micro-evidence on individuals' asset allocation*, Cambridge, MA: National Bureau of Economic Research, Working Paper 12809

Burchardt, T. (2004), 'Selective inclusion', in Stewart, K. and Hills, J. (eds.), *A more equal society?* Bristol: Policy Press

Burgoyne, C. (1995), 'Financial organization and decision-making within Western "households" ', *Journal of Economic Psychology*, 16: 421–430

Burrell, K. (2011), 'Going steerage on Ryanair: Cultures of migrant air travel between Poland and the UK', *Journal of Transport Geography*, 19: 1023–1030

Butler, J. (1990), *Gender trouble: Feminism and the subversion of identity*, New York: Routledge

Buzan, B., Wæver, O. and de Wilde, J. (1998), *Security: A new framework for analysis*, Boulder, CO: Lynne Rienner

Byrnes J. P., Miller, D. C. and Schafer W. D. (1999), 'Gender differences in risk taking: A meta-analysis', *Psychological Bulletin*, 125: 367–383

Cabinet Office (2002), *Risk: Improving government's capability to handle risk and uncertainty*, London: Cabinet Office, Strategy Unit Report

Camerer, C., Babcock, L., Loewenstein, G. and Thaler, R. H. (1997), 'Labor supply of New York City cab drivers: One day at a time', *Quarterly Journal of Economics*, 112: 407–441

Camerer, C. and Weber, M. (1992), 'Recent developments in modelling preferences: Uncertainty and ambiguity', *Journal of Risk and Uncertainty*, 5: 325–370

Campbell, S. (2005), 'Determining overall risk', *Journal of Risk Research*, 8: 569–581

Carling, J. (2007), 'Migration control and migrant fatalities at the Spanish-African borders', *International Migration Review*, 41: 316–343

Carpenter, Daniel P. (2001), *The forging of bureaucratic autonomy: Reputations, networks, and policy innovation in executive agencies, 1862–1928*, Princeton: Princeton University Press

Carrington, W. J, Detragiache, E. and Vishwanath, T. (1996), 'Migration with endogenous moving costs', *The American Economic Review*, 86: 909–930

Carretero, M. H. (2008), *Risk taking in unauthorized migration*, Master's thesis, Tromsø: University of Tromsø, Social Science Faculty

Casselman, B. (2013), 'Risk-averse culture infects U.S. workers, entrepreneurs', *The Wall Street Journal*, 2 June 2013. Available at: online.wsj.com/article/SB1000142 4127887324031404578481162903760052.html [Accessed October 2013]

'CBI and Boris Johnson attack immigration policy' (2012, November 16) *Daily Telegraph*, Available at: www.telegraph.co.uk/finance/economics/9682727/CBI-and-Boris-Johnson-attack-immigration-policy.html

Cesarini, D., Dawes, C. T., Johannesson, M., Lichtenstein, P. and Wallace, B. (2009), 'Genetic variation in preferences for giving and risk-taking', *Quarterly Journal of Economics*, 124: 809–842

Ceyhan, A. and Tsoukala, A. (2002), 'The securitization of migration in western societies: Ambivalent discourses and policies', *Alternatives: Global, Local, Political*, 27 (Special Issue): 21–39

Chang, C. C., DeVaney, S. A and Chiremba, S. T. (2004), 'Determinants of subjective and objective risk tolerance', *Journal of Personal Finance*, 3: 53–66

Chiswick, B. R. (1978), 'The effect of Americanization on the earnings of foreign-born men', *The Journal of Political Economy*, 86: 897–892

Chiswick, B. R. (1980), 'The earnings of white and coloured male immigrants in Britain', *Economica*, 47: 81–87

Chiswick, B. R. (2008), 'Are immigrants favourably self-selected? An economic analysis', in Brettell, C. B. and Hollifield, J. (eds.), *Migration theory: Talking across disciplines* (2nd ed.), New York: Routledge, pp. 63–82

Chiswick, B. R. and Hatton, T. J. (2003), 'International migration and the integration of labor markets', in Bordo, M., Taylor, A. and Williamson, J. (eds.), *Globalization in historical perspective*, Chicago: National Bureau of Economic Research Conference Reports, University of Chicago Press, pp. 65–119

Chiswick, B. R. and Miller, P. M. (1996), 'Ethnic networks and language proficiency among immigrants', *Journal of Population Economics*, 9: 19–35

Choldin, H. M. (1973), 'Kinship networks in the migration process', *Demography*, 10: 163–175

Church, J. and King, I. (1993), 'Bilingualism and network externalities', *Canadian Journal of Economics*, 26: 337–345

Clark, W. A. V. and Withers, S. D. (2002), 'Disentangling the interaction of migration, mobility, and labour-force participation', *Environment and Planning A*, 34: 923–945

Cohen, J. H. (2004), *The culture of migration in southern Mexico*, Austin: University of Texas Press

Collyer, M. (2010), 'Stranded migrants and the fragmented journey', *Journal of Refugee Studies*, 23: 273–290

Collyer, M. and de Haas, H. (2012), 'Developing dynamic categorisations of transit migration', *Population, Space and Place*, 18: 468–481

Coombs, G. (1978), 'Opportunities, information networks and the migration distance relationship', *Social Networks*, 1: 257–276

Corno, L. and de Walque, D. (2012), *Mines, migration and HIV/AIDS in southern Africa*, Washington D. C.: World Bank, Policy Research Working Paper No. 5966

Crang, M. (2004), 'Cultural geographies of tourism', in Lew, A. A., Hall, C. M. and Williams, A. M. (eds.), *A companion to tourism*, Oxford: Wiley-Blackwell Publishing, pp. 74–84

Cunliffe, S. K. (2006), 'Best education network think tank v keynote address: Risk management for tourism: Origins and needs', *Tourism Review International*, 10: 27–38

DaVanzo, J. (1981), 'Repeat migration, information costs and location-specific capital', *Population and Environment*, 4: 45–73

DaVanzo, J. (1983), 'Repeat migration in the United States: Who moves back and who moves on?', *The Review of Economics and Statistics*, 65: 552–559

Dake, D. (1992), 'Myths of nature: Culture and the social construction of risk', *Journal of Social Issues*, 48: 21–27

Dake, K. (1991), 'Orienting dispositions in the perception of risk: An analysis of contemporary worldviews and cultural biases', *Journal of Cross-Cultural Psychology*, 22: 61–82

van Dalen, H. P. and Henkens, K. (2012), 'Explaining low international labour mobility: The role of networks, personality, and perceived labour market opportunities', *Population, Space Place*, 18: 31–44

Daruvala, D. (2007), 'Gender, risk and stereotypes', *Journal of Risk and Uncertainty*, 35: 265–283

Dean, M. (1999), *Governmentality, power and rule in modern society*, London: Sage

Debreu, G. (1959), *The theory of value: An axiomatic analysis of economic equilibrium*, New Haven and London: Yale University Press

Deleuze, G. (1995), 'Postscript on control societies', in Deleuze, G. (ed.), *Negotiations, 1972–1990*, New York: Columbia University Press

Delgado, R. and Stefancic, J. (1992), 'Images of the outsider in American law and culture: Can free expression remedy systemic social ills?', *Cornell Law Review*, 77: 1258–1281

Dench, S., Hurstfield, J., Hill, D. and Akroyd, K. (2006), *Employers' use of migrant labour: Main report*, London: Home Office, Online Report 04/06. www.northamptonshireobservatory.co.uk/docs/docrdsolr0406060818113351.pdf

Denzin, N. (1991), *Images of postmodern society: Social theory and contemporary cinema*, London: Sage

Derose, K. P., Escarce, J. J. and Lurie, N. (2007), 'Immigrants and health care: Sources of vulnerability', *Health Affairs*, 26: 1258–1268

Diamond, J. (1997), *Guns, germs and steel: A short history of everybody for the last 13,000 years*, London: Vintage

Dingwall, R. (1999), 'Risk society: The cult of theory and the millennium?', *Social Policy & Administration*, 33: 474–491

Dohmen, T., Falk, A., Huffman, D. and Sunde, U. (2006), *The intergenerational transmission of risk and trust attitudes*, Bonn: Institute for the Study of Labour, IZA Discussion Paper 2380

Dohmen, T., Falk, A., Huffman, D., Sunde, U., Schupp, J. and Wagner, G. G. (2005), *Individual risk attitudes: New evidence from a large, representative, experimentally-validated survey*, Bonn: Institute for the Study of Labour, IZA Discussion Paper 1730

Donkers, B., Melenberg, B. and van Soest, A. (2001), 'Estimating risk attitudes using lotteries: A large sample approach', *Journal of Risk and Uncertainty* 22: 165–195

Dorfles, P. (1991), *Guardando all'Italia: Influenza delle TV e delle Radio Italiane sull'Esodo degli Albanesi*. Rome: RAI-VQPT.

Douglas, M. (1985), *Risk acceptability according to the social sciences*, New York: Sage

Douglas, M. (1992), *Risk and blame: Essays in cultural theory*, Routledge: London

Douglas, M. and Wildavsky, A. (1982), *Risk and culture: An essay on the selection of technological and environmental dangers*, Berkeley: University of California Press

Doyle, P. and Hutchinson, P. (1973), 'Individual differences in family decision making', *Journal of the Market Research Society*, 15: 193–206

Dustmann, C. (1997), 'Return migration, uncertainty and precautionary savings', *Journal of Development Economics*, 52: 295–316

Dustmann, C. and Weiss, Y. (2007), 'Return migration: Theory and empirical evidence from the UK', *British Journal of Industrial Relations*, 45: 236–256

ECRE (2007), *Defending refugees' access to protection in Europe*, Brussels: European Council on Refugees and Exiles

Edelman, M. (1988), *Construction of the political spectacle*, Chicago: University of Chicago Press

Edkins, J. and Pin-Fat, V. (2004), 'Introduction: Life, power and resistance', in Edkins, J., Pin-Fat, V. and Shapiro, M. J. (eds.), *Sovereign lives: Power in global politics*, London: Routledge, pp. 1–22

Edwards, A. (2009), 'Crime, terror and immigration as one global risk?', *Risk and Regulation Magazine*, 17: 8–9

Eisenberg, D. T. A., Campbell, B., Gray, P. B. and Sorenson, M. D. (2008), 'Dopamine receptor genetic polymorphisms and body composition in undernourished pastoralists: An exploration of nutrition indices among nomadic and recently settled Ariaal men of northern Kenya', *BMC Evolutionary Biology*, 8:173

Ellsberg, D. (1961), 'Risk, ambiguity, and the savage axioms', *Quarterly Journal of Economics*, 75: 643–669

Ellsberg, D. (2001). *Risk, ambiguity, and decision*, New York: Garland Publishing

Elsrud, T. (2001), 'Risk creation in traveling: Backpacker adventure narration', *Annals of Tourism Research*, 28: 597–617

Engel, J. F., Blackwell, R. D. and Miniard, P. W. (1990), *Consumer behavior*, Chicago: Dryden Press

Engel, U. and Strasser, H. (1998), 'Global risks and social inequality: Critical remarks on the risk-society hypothesis', *Canadian Journal of Sociology*, 23: 91–103

Epstein, G. S. (2008), 'Herd and network effects in migration decision-making', *Journal of Ethnic and Migration Studies*, 34: 567–583

Erasmus, A. C., Boshoff, E. and Rousseau, G. G. (2001), 'Consumer decision-making models within the discipline of consumer science: A critical approach', *Journal of Family Ecology and Consumer Sciences*, 29: 82–90

Ericson, R. V. and Haggerty, K. (1997), *Policing the risk society*, Toronto: University of Toronto Press

Espinosa, K. E. and Massey, D. S. (1999), 'Undocumented migration and the quantity and quality of social capital', in Pries, L. (ed.), *Migration and transnational spaces*, Aldershot: Ashgate, pp. 106–137

Etkin, D. (1999), 'Risk transference and related trends: driving forces towards more mega-disasters', *Environmental Hazards*, 1: 69–75

Ewald, F. (1986), *L' état providence*, Paris: Grasset

Fafchamps, M. (1992), 'Solidarity networks in preindustrial societies: Rational peasants with a moral economy,' *Economic Development and Cultural Change*, 41: 147–174

Faist, T. (1997), 'The crucial meso-level', in Hammar, T., Brochmann, G., Tamas, K. and Faist, T. (eds.), *International migration, immobility and development*, multidisciplinary perspectives, Oxford: Berg, pp. 187–217

Fassmann, H. and Hintermann, C. (1998), 'Potential east-west migration, demographic structure, motives and intentions', *Czech Sociological Review*, 6: 59–72

Fawcett, J. T. (1989), 'Networks, linkages, and migration systems', *International Migration Review*, 23: 671–680

Fehr, E., Fischbacher, U., Naef, M., Schupp, J. and Wagner, G. G. (2006), *A comparison of risk attitudes in Germany and the U.S.*, Mimeo, Zurich: Institute for Empirical Research in Economics, University of Zurich

Finlayson, A. (2003), *Making sense of new labour*, London: Lawrence & Wishart

Fischer, P. A., Martin, R. and Straubhaar, T. (1997), 'Should I stay or should I go', in Hammar, T., Brochmann, G., Tams, K. and Faist, T. (eds.), *International migration, mobility and development, multidisciplinary perspectives*, Oxford: Berg, pp. 49–90

Fishburn, P. C. (1974), 'Lexicographic orders, utilities and decision rules: A survey', *Management Science*, 20: 1442–1471

Flynn, D. (2005), 'On new borders, new management: The dilemmas of modern immigration policies', *Ethnic and Racial Studies*, 28: 463–490

Folbre, N. (1986), 'Cleaning house: New perspectives on households and economic development', *Journal of Development Economics*, 22: 5–40

Foucault, M. (1980), 'The confessions of the flesh', in Gordon, C. (ed.), *Power/Knowledge: Selected interviews and other writings, 1972–1977*, New York: Pantheon Books

Foucault, M. (1991), 'Governmentality', in Burchell, G., Gordon, C., and Miller, P. (eds.), *The Foucault effect*, Chicago: University of Chicago Press, pp. 87–104

Foucault, M. (2007), *Security, terrorism, population: Lectures at the Collège de France*, Basingstoke: Palgrave

Fox, C. R. and Tversky, A. (1995), 'Ambiguity aversion and comparative ignorance', *The Quarterly Journal of Economics*, 110: 585–603

Fox, S. C. (1988), 'General John DeWitt and the proposed internment of German and Italian aliens during World War II', *Pacific Historical Review*, 57: 407–438

Fraser, N. (1989), 'Talking about needs: Interpretive contests as political conflicts in welfare-state societies', *Ethics*, 99: 291–313

Freeman, G. P. (1986), 'Migration and the political economy of the welfare state', *Annals of the American Academy of Political and Social Science*, 485: 51–63

Friend, I. and Blume, M. E. (1975), 'The demand for risky assets', *American Economic Review*, 65: 900–922

Furedi, F. (2006), *Culture of fear revisited*, London: Cassell

Furlong, A. and Cartmel, F. (1997), Young people and social change: Individualisation and risk in late modernity, Buckingham: Open University Press

Galor, O. and Stark, O. (1991), 'The probability of return migration, migrants' work effort and migrants' performance', *Journal of Development Economics*, 35: 399–405

Gamburd, M. R. (2000), *The kitchen spoon's handle: Transnationalism and Sri Lanka's migrant housemaids*, Ithaca: Cornell University Press

Ganzach, Y. and Krantz, D. H. (1990), 'The psychology of moderate prediction I: Experience with multiple determination', *Organizational Behavior & Human Decision Processes*, 47: 177–204

García-Retamero, R. and Rieskamp, J. (2008), 'Adaptive mechanisms for treating missing information: A simulation study', *The Psychological Record*, 58: 547–568

Ghosh, B.(1985), 'Brain migration from the third world: An implicative analysis', *Rivista Internazionale di Scienze Economiche e Commerciali*, 34: 21–34

Ghosh, B. (ed.) (2000), *Managing Migration: Time for a new international regime*, Oxford: Oxford University Press

Giddens, A. (1990), *Consequences of modernity*, Cambridge: Polity Press

Giddens, A. (1991), *Modernity and self-identity: Self and society in the late modern age*, Cambridge: Polity Press

Giddens, A. (1999), 'Risk and responsibility', *Modern Law Review*, 62: 1–10

de Goede, M. (2003), 'Hawala discourses and the war on terrorist finance', *Environment and Planning D: Society and Space*, 21: 513–532

Goffart, W. A. (2006), *Barbarian tides: The migration age and the later Roman Empire*, Philadelphia: University of Pennsylvania Press

Goffman, E. (1967), *Interaction ritual*, New York: Anchor

Gonzales, R. and Wu, G. (1999), 'On the shape of the probability weighting function', *Cognitive Psychology*, 38: 129–166

Graham, B. and Shaw, J. (2008), 'Low-cost airlines in Europe: Reconciling liberalization and sustainability', *Geoforum*, 39: 1439–1451

Graham, J. D. and Weiner, J. B. (eds.) (1995), *Risk versus risk: Tradeoffs in protecting health and the environment*, Cambridge: Harvard University Press

Green, A. E., Power, M. R. and Jang, D. M. (2008), 'Trans-Tasman migration: New Zealanders' explanations for their move', *New Zealand Geographer*, 64: 34–45

Green, J. (1997), 'Risk and the construction of social identity: Children's talk about accidents', *Sociology of Health & Illness*, 19: 457–479

Green, T. and Winters, L. A. (2012), 'Economic crises and migration: Learning from the past and the present', (Chapter 3), in Sirkeci, I., Cohen, J. H. and Ratha, D. (eds.), *Remittances during the global financial crisis and beyond*, Washington D.C.: The World Bank, pp. 35–52

Grönqvist, H. (2006), 'Ethnic enclaves and the attainment of immigrant children', *European Sociological Review*, 22: 369–382

Gupta, A. and Ferguson, J. (1997), 'Culture, power, place: ethnography at the end of an era', in Gupta, A. and Ferguson, J. (eds.), *Culture, power, place: Explorations in critical anthropology*, Durham, NC: Duke University Press, pp. 1–29

Gushulak, B. D. and MacPherson, D. W. (2004), 'Globalization of infectious diseases: The impact of migration', *Clinical Infectious Diseases*, 38: 1742–1748

Haan, A. de, Brock, K., Carswell, G., Coulibaly, N., Seba, H. and Ali Toufique, K. (2000), *Migration and livelihoods: Case studies in Bangladesh, Ethiopia and Mali*, Brighton: IDS, University of Sussex, Institute of Development Studies, Research Report 46

Haas, H. de (2008), *Irregular migration from West Africa to the Maghreb and the European Union: An overview of recent trends*, Geneva: International Organization for Migration, IOM Research Series No. 32

Haas, L. (1980), 'Role sharing couples: A study of egalitarian marriages', *Family Relations*, 29: 289–296

Hajduchova, A. (2003), 'Práca au pair' (au pair work), *SME Daily, Supplement*, 18 June 2003, pp. 30–31

Halek, M. and Eisenhauer, J. G. (2001), 'Demography of risk aversion', *Journal of Risk and Insurance*, 68: 1–24

Hallahan T. A., Faff, R. W. and McKenzie, M. D. (2004), 'An empirical investigation of personal financial risk tolerance', *Financial Services Review*, 13: 57–78

Halliday, T. (2006), 'Migration, risk, and constraints in El Salvador', Economic Development and Cultural Change, 54: 893–925

Hamid, H. (2007), 'Transnational migration: Addressing or importing risk', Paper presented at the conference on international migration, multi-local livelihoods and human security: Perspectives from Europe, Asia and Africa, Netherlands: Institute of Social Studies, 30 and 31 August 2007

Hamilton, G. G. (1978), 'The structural sources of adventurism: The case of the California gold rush', American Journal of Sociology, 83: 1466–1490

Hansen, R. (2000), *Immigration and citizenship in postwar Britain*, Oxford: Oxford University Press

Harbeck, M., Seifert., L., Hänsch, S., Wagner, D. M., Birdsell, D., Parise, K., Wiechmann, I., Gisela, G., Astrid, T., Lothar Z., Riehm, J., and Holger C. (2013), 'Yersinia pestis DNA from skeletal remains from the 6th century AD reveals insights into Justinianic plague', *PLoS Pathog* 9: e1003349 doi:10.1371/journal. ppat.1003349

Harper, K. N., Ocampo, P. S., Steiner, B. M., George, R. W., Silverman, M. S. et al. (2008), 'On the origin of the Treponematoses: A phylogenetic approach', *PLoS Neglected Tropical Diseases*, 2: e148 doi:10.1371/journal.pntd.0000148

Harris, J. R. and Todaro, M. P. (1970), 'Migration, unemployment and development: A two sector analysis', *American Economic Review*, 60: 126–142

Hartog, J., Ferrer-I-Carbonell, A. and Jonker, N. (2000), *On a simple measure of individual risk aversion*, Munich: CESifo, Working Paper 363

Hassan, I. (1985), 'The culture of postmodernism', *Theory, Culture and Society*, 2: 119–132. doi: 10.1177/0263276485002003010

Haug, S. (2008), 'Migration networks and migration decision-making', *Journal of Ethnic and Migration Studies*, 34: 585–605

Hayenhjelm, M. (2006), 'Out of the ashes: Hope and vulnerability as explanatory factors in individual risk taking', *Journal of Risk Research*, 9: 189–204

Heath, C. and Tversky, A. (1991), 'Preference and belief: Ambiguity and competence in choice under uncertainty', *Journal of Risk and Uncertainty*, 4: 2–28

Heitmueller, A. (2005), 'Unemployment benefits, risk aversion, and migration incentives', *Journal of Population Economics*, 18: 93–112

Held, D. (ed.) (2000), *A globalizing world? Culture, economics and politics*, London: Routledge

Herman, E. (2006), 'Migration as a family business: The role of personal networks in the mobility phase of migration', *International Migration*, 44: 191–230

Hickman, M., Mai, N. and Crowley, H. (2012), *Migration and social cohesion in the UK*, Basingstoke: Palgrave Macmillan

Himmelweit, S. (1998), 'Decision making in households', in Himmelweit, S. et al. (eds.), *Understanding economic behaviour: Households*, Open University, Milton Keynes, pp. 181–219

Hoddinott, J. (1992), 'Modelling remittance flows in Kenya', *Journal of African Economies*, 1: 206–232

Hoffmann-Nowotny, H. J. (1981), 'A sociological approach towards general theory of migration', in Kritz, M. M., Keely, Ch. B. and Tomasi, S. M. (eds.), *Global trends in migration: Theory and research on international population movements*, New York: The Centre of Migration Studies, pp. 64–83

Hollifield, J. F. (2008), 'The politics of international migration', in Brettell, C. B. and Hollifield, J. F. (eds.), *Migration theory: Talking across disciplines*, New York: Routledge

Home Office (1998) *Fairer, faster and firmer: A modern approach to immigration and asylum*, London: UK Home Office White Paper, Cm4018, Norwich: Stationery Office

House of Commons Home Affairs Committee (2008), *Bulgarian and Romanian accession to the EU: Twelve months on*. London: The Stationery Office, Second Report of Session 2007–08

Hughes, T. P. (2004), *Human-built world: How to think about technology and culture*, Chicago: University of Chicago Press

Huntingdon, S. P. (2004), *Who are we? The challenge to America's identity*, New York: Simon and Schuster

Huysmans, J. (2000), 'The European Union and the securitization of migration', *Journal of Common Market Studies*, 38: 751–777

IASC (2011), *IASC guidelines on the protection of persons in situations of natural disasters*, Washington D.C.: Brookings-Bern Project on Internal Displacement

Ibrahim, M. (2005), 'The securitization of migration: A racial discourse', *International Migration*, 43: 163–187

ILO (2005), *A global alliance against forced labour*, Geneva: International Organization for Migration

IOM International Organisation for Migration (2008), *Migration and climate change*, prepared for IOM by Brown, O., Geneva: International Organization for Migration, IOM Migration Research Series No. 31

ISO (2002), *Risk management vocabulary*, ISO/IEC Guide 73. Geneva: ISO

Jaccard, J. and Wood, G. (1988), 'The effects of incomplete information on the formation of attitudes toward behavioral alternatives', *Journal of Personality and Social Psychology*, 54: 580–591

Jaeger, D. A., Bonin, H., Falk, A., Huffman, D., Dohmen, T. and Sunde, U. (2007), *Direct evidence on risk attitudes and migration*, Bonn: Institute for the Study of Labour, Discussion Paper No. 2655

Jennings, W. (2007), *At no serious risk? Border control and asylum policy in Britain, 1994–2004*, London: London School of Economics, Centre for Analysis of Risk and Regulation, Discussion Paper No. 39

Johnson, R. D. and Levin, I. P. (1985), 'More than meets the eye: The effect of missing information on purchase evaluations', *Journal of Consumer Research*, 12: 169–177

Jones, R. C. (1989), 'Causes of Salvadoran migration to the United States', *The Geographical Review*, 79: 183–194

Jones, R. (2012), *Border walls: Security and the war on terror in the United States, India and Israel*, London: Zed Books

de Jong, G. F., Abad, R. G., Arnold, F., Carino, B. V., Fawcett, J. T. and Gardner, R. W. (1983), 'International and internal migration decision making: A value-expectancy based analytical framework of intentions to move from a rural Philippine province', *International Migration Review*, 17: 470–484

Jourard, S. M. (1971), *Self-disclosure: An experimental analysis of the transparent self*, New York: Wiley

Kahneman, D. (1994), 'New challenges to the rationality assumption', *Journal of Institutional and Theoretical Economics/Zeitschrift für die gesamte Staatswissenschaft*, 150: 18–36

Kahneman, D., Thaler, R. H. (2006), 'Utility maximization and experienced utility', *Journal of Economic Perspectives*, 20: 221–234

Kahneman, D. and Tversky, A. (1979), 'Prospect theory: An analysis of decision under risk', *Econometrica*, 47: 263–292

Kaplan, S. (1991), 'Risk assessment and risk management—Basic concepts and terminology', in Knief, R., Briant, V. B., Lee, R. B., Long R. L., and Mahn, J. A. (eds.) *Risk management: Expanding horizons in nuclear power and other industries*, Boston, MA: Hemisphere Publ. Corp., pp. 11–28

Kapur, D. (2004), *Remittances: The new development mantra*, Research papers for the Intergovernmental Group of Twenty Four on International Monetary Affairs, United Nations Conference on Trade and Development, Discussion Paper Series No. 29, New York and Geneva: United Nations

Kasperson, R. E. (1992), 'The social amplification of risk: Progress in developing an integrative framework', (Chapter 6), in Krimsky, S. and Golding, D. (eds.), *Social theories of risk*, Connecticut: Praeger, pp. 153–178

Kasperson, R. E., Renn, O., Slovic, P., Brown, H. S., Emel, J., Goble, R., Kasperson, J. X., and Ratick, S. (1988), 'The social amplification of risk: A conceptual framework', *Risk Analysis*, 8: 177–187

Katz, E., and Stark, O. (1986), 'Labor migration and risk aversion in less developed countries', *Journal of Labor Economics*, 4: 134–149

Kemefick, E. (2013), 'Health coverage for undocumented migrants: A European smorgasbord', Chicago Policy Review 6 March 2013. Available at: chicago policyreview.org/2013/03/06/health-coverage-for-undocumented-migrants-a-european-smorgasbord/ [Accessed November 2013]

Kilka, M. and Weber, M. (2001), 'What determines the shape of the probability weighting function under uncertainty?', *Management Science*, 47: 1712–1726

King, D. (2005), *The liberty of strangers: Making the American nation*, New York: Oxford University Press

King R. and Ruiz-Gelices, E. (2003), 'International student migration and the European "year abroad": Effects on European identity and subsequent migration behaviour', *International Journal of Population Geography*, 9: 229–252

King, R. and Skeldon, R. (2010), '"Mind the gap!" Integrating approaches to internal and international migration', *Journal of Ethnic and Migration Studies*, 36: 1619–1646

King, R., Warnes, A. and Williams, A. M. (2000), *Sunset lives: British retirement migration to the Mediterranean*, Oxford: Berg

Kivetz, R., Netzer, O. and Schrift, R. (2008), 'The synthesis of preference: Bridging behavioural decision research and marketing science', *Journal of Consumer Psychology*, 18: 179–186

Kivetz, R. and Simonson, I. (2000), 'The effects of incomplete information on consumer choice', *Journal of Marketing Research*, 37: 427–448

Knight, F. H. (1921), *Risk, uncertainty and profit*, Boston: Houghton Mifflin

Kofman, E. (2002), 'Contemporary European migrations, civic stratification and citizenship', *Political Geography*, 21: 1035–1054

Kofman, E. (2004), 'Family-related migration: Critical review of European Studies', *Journal of Ethnic and Migration Studies*, 30: 243–263

Koser, K. (2005), *Irregular migration, state security and human security*, London: Global Commission on International Migration (GCIM)

Koser, K. (2008), 'Why migrant smuggling pays', *International Migration*, 46: 3–26Koslowski, R. (2001), 'Personal security and state sovereignty in a uniting Europe', in Guiraudon, V. and Joppke, C. (eds.), *Controlling a new migration world*, New York: Routledge, pp. 99–120

Krantz, D. H. and Kunreuther, H. C. (2007), 'Goals and plans in decision making', *Judgment and Decision Making*, 2: 137–168

Ku, L. (2006), 'Why immigrants lack adequate access to health care and health insurance', Migration Information Source. Available at: www.migrationinformation.org/usfocus/display.cfm?ID=417 [Accessed June 2013]

Ku, L. and Matani, S. (2001), 'Left out: Immigrants' access to health care and insurance', *Health Affairs*, 20: 247–256

Kunreuther, H., Meyer, R., Zeckhauser, R., Slovic, P., Schwartz, B., Schade, C., Luce, M. F., Lippman, S., Krantz, D., Kahn, D. and Hogarth, R. (2002), 'High stakes decision making: Normative, descriptive and prescriptive considerations', *Marketing Letters*, 13: 259–268

Lackman, C. and Lanasa, J. M. (1993), 'Family decision-making theory: An overview and assessment', *Psychology & Marketing* 10: 81–93

Lahav, G. and Guiraudon, V. (2006), 'Actors and venues in immigration control: Closing the gap between political demands and policy outcomes', *West European Politics*, 29: 201–223

Lash, S. (2000), 'Risk culture', in Adam, B., Beck, U. and van Loon, J. (eds.), *The risk society and beyond: Critical issues for social theory*, Sage: London, pp. 47–62

Lebel, L., Xu, J., Bastakoti, R. C. and Lamba, A. (2010), 'Pursuits of adaptiveness in the shared rivers of Monsoon Asia', *International Environmental Agreements: Politics, Law and Economics*, 10: 355–375

Lee, B. K. and Lee, W. N. (2004), 'The effect of information overload on consumer choice quality in an on-line environment', *Psychology and Marketing*, 21: 159–183

Lee, E. S. (1966), 'A theory of migration', *Demography*, 3: 45–47

Leiken, R. S. (2004), *Bearers of global jihad? Immigration and national security after 9/11*, Washington, D. C.: Nixon Center, p. 6. (Gunaratna's quotation is attributed to his presentation at the Nixon Center on 1December 2003)

Levi, M. and Wall, D. (2004), 'Technologies, security and privacy in the post 9/11 European information society', *Journal of Law and Society*, 31: 194–220

Levin, I. P., Johnson, R. D., Ruso, C. P. and Deldin, P. J. (1985), 'Framing effects in judgment tasks with varying amounts of information', *Organizational Behavior and Human Decision Processes*, 36: 362–377

Lewis, J. D. and Weigert, A. (1985), 'Trust as a social reality', *Social Forces*, 63: 967–985

Lewis, W. A. (1954), 'Economic development with unlimited supplies of labor', *The Manchester School of Economic and Social Studies*, 22: 139–191

Lillard, L. and Willis, R. (1997), 'Motives for intergenerational transfers: Evidence from Malaysia', *Demography*, 34: 115–134

Litwak, E. and Szelenyi, I. (1969), 'Primary group structures and their functions: Kin, neighbors, and friends', *American Sociological Review*, 34: 465–481

Long, L. H. (1975), 'Does migration interfere with children's progress in schools', *Sociology Educ,* 48: 369–381

Lovejoy, P. E. (1989), 'The impact of the Atlantic slave trade on Africa: A review of the literature', *The Journal of African History*, 30: 365–394

Lowell, B. L. and Findlay, A. (2002), *Migration of highly skilled persons from developing countries: Impact and policy responses*, Geneva: International Labor Office

Lowrance, W. (1976), *Of acceptable risk—Science and the determination of safety*, Los Altos, CA: William Kaufmann Inc.

Lucas, R. E. B. and Stark, O. (1985), 'Motivations, to remit: Evidence from Botswana', *Journal of Political Economy*, 93: 901–918

Lupton, D. (2006), 'Sociology and risk', in Mythen, G. and Walkate, S. (eds.), *Beyond the risk society: Critical reflections on risk and human security*, Maidenhead, England: Open University Press, pp. 11–24

Lyng, S. (2008), 'Edgework, risk and uncertainty', in Zinn J. O. (ed.), *Social theories of risk and uncertainty: An introduction*, Blackwell: Oxford, pp. 106–137

Macgill, S. M. and Siu, Y. L. (2005), 'A new paradigm for risk analysis', *Futures*, 37: 1105–1131

Mai, N. (2004), 'Looking for a more modern life…: The role of Italian television in the Albanian migration to Italy', *Westminster papers in communication and culture*, London: University of Westminster, 1: 3–22

Mai. N. (2011), 'Tampering with the sex of 'Angels': Migrant male minors and young adults selling sex in the EU', *Journal of Ethnic and Migration Studies*, 37(8): 1237–1252

Maier, G. (1985), 'Cumulative causation and selectivity in labour market oriented migration caused by imperfect information', *Regional Studies*, 19: 231–241

Malmberg, G. (1997), 'Time and space in international migration', in Hammar, T., Brochmann, G., Tams, K. and Faist, T. (eds.), *International migration, mobility and development, multidisciplinary perspectives*, Oxford: Berg, pp. 21–48

Mandel, R. (1990), 'Shifting centers and emergent identities: Turkey and Germany in the lives of Turkish gastarbeiter', in Eickelman, D. and Piscatori, J. (eds.), *Muslim travelers: Pilgrimage, migration and the religious imagination*, Berkeley: University of California Press, pp. 153–171

Manning, P. (2005), *Migration in world history*, New York and London: Routledge

Marr, W. L. and Paterson D. G. (1980), *Canada: An economic history*, Toronto: Macmillan of Canada

Massey, D. S. (1990), 'Social structure, household strategies, and the cumulative causation of migration', *Population Index*, 56: 3–26

Massey, D. S. (1998), 'March of folly: US immigration policy after NAFTA', *The American Prospect*, 9:37, 22–33

Massey, D. S., Arango, J., Hugo, G., Kouaouci, A., Pellegrino, A. and Taylor, J. E. (1993), 'Theories of international migration: A review and appraisal', *Population and Development Review*, 19: 431–466

Massey, D. S., Arango, J., Hugo, G., Kouaouci, A., Pellegrino, A. and Taylor, J. E. (1994), 'An evaluation of international migration theory: The North American case', *Population and Development Review*, 20: 699–751

Massey, D. S. and Denton, N. A. (1989), 'Hypersegregation in US metropolitan areas: Black and Hispanic segregation along five dimensions', *Demography*, 26: 373–391

Masuda, J. R. and Garvin, T. (2006), 'Place, culture, and the social amplification of risk', *Risk Analysis*, 26: 437–454

Matten, D. (2004), 'The impact of the risk society thesis on environmental politics and management in a globalising economy: Principles, proficiency, perspectives', Journal of Risk Research, 7: 377–398

Matthews, L. J., and P. M. Butler. (2011), 'Novelty-seeking DRD4 polymorphisms are associated with human migration distance out-of-Africa after controlling for neutral population gene structure', *American Journal of Physical Anthropology*, 145: 382–389.

McCourt, F. (1996), *Angela's ashes*, New York: Scribner

McCusker, R. (2005), 'Underground banking: Legitimate remittance network or money laundering system?', *Trends and issues in crime and criminal justice*, No. 300, Australian Institute of Criminology

McGregor, J., Marazzi, L., Mpofu, B., (2011), *Conflict, migration and the environment: The case of Zimbabwe*. The Government Office for Science, London Online. Available at: www.bis.gov.uk/foresight/migration [Accessed October 2012]

McKenzie, D., Gibson, J. and Stillman, S. (2013), 'A land of milk and honey with streets paved with gold: Do emigrants have over-optimistic expectations about incomes abroad?', *Journal of Development Economics*, 102: 116–127.

McKenzie, D. and Rapoport, H. (2010), 'Self-selection patterns in Mexico-U.S. migration', *The Review of Economics and Statistics*, 92: 811–821

McLaren, L. and Johnson, M. (2004), 'Understanding the rising tide of anti-immigrant sentiment' in *British Social Attitudes: The 21st Report*, London: National Centre for Social Research/Sage

McLeman, R. (2011), 'Settlement abandonment in the context of global environmental change', *Global Environmental Change*, 21: S108–S120

McLeman, R. and Smit, B. (2006), 'Migration as an adaptation to climate change', *Climatic Change*, 76: 31–53

Médecins Sans Frontières (MSF) (2005), *Violence et immigration. Rapport sur l'immigration d'origine subsaharienne (ISS) en situation irrégulieré au Maroc*, Barcelona: MSF

Mehta, J. (2007), *Being economic: Perspectives on risk and rationality*, Canterbury: University of Kent at Canterbury, SCARR Working Paper 2007/18

Melendez, E. (1994), 'Puerto Rican migration and occupational selectivity, 1982–88', *International Migration Review*, 28: 49–67

Meyer, J. B. (2001), 'Network approach versus brain drain: Lessons from the diaspora', *International Migration*, 39: 91–110

Migration Advisory Committee (2011), *Skilled shortage sensible: Full review of the recommended shortage occupation lists for the UK and Scotland*, Croydon: Migration Advisory Committee

Miller, D. and Paulson, A. (1999), *Informal insurance and moral hazard: Gambling and remittances in Thailand*, Manuscript, Northwestern University

Mincer, J. (1978), 'Family migration decisions', *Journal of Political Economy*, 86: 769–773

Mohapatra, S. and Ratha, D. (2012), 'Forecasting migrant remittances during the global financial crisis', (Chapter 2), in Sirkeci, I., Cohen, J. H. and Ratha, D. (eds.), *Remittances during the global financial crisis and beyond*, Washington D. C.: The World Bank, pp. 23–34

Momsen, J. H. (1999), 'Maids on the move: victim or victor', in Hensall-Momsen, J. (ed.), *Gender, migration and domestic service*, London; New York: Routledge, pp.1–20

Monti, M., Martignon, L., Gigerenzer, G. and Berg, N. (2009), 'The impact of simplicity on financial decision-making', *Proceedings of the 31st Annual Conference of the Cognitive Science Society*, 386: 1846–1851

Morris, L. (2002), *Managing migration: Civic stratification and migrants' rights*, London: Routledge

Myers, N. (2002), 'Environmental refugees: a growing phenomenon of the 21st century', *Philosophical Transactions of the Royal Society*, London: B 357 (1420), 609–613

Neske, M. (2006), 'Human smuggling to and through Germany', *International Migration*, 44: 121–64

von Neumann, J. and Morgenstern, O. (1944), *Theory of games and economic behavior*, Princeton: Princeton University Press

Neumayer, E. and Barthel, F. (2011), 'Normalizing economic loss from natural disasters: A global analysis', *Global Environmental Change*, 21: 13–24

Newland, K. (2011), *Climate Change and Migration Dynamics*, Washington DC: Migration Policy Institute

Nickels, H. C., Thomas L., Hickman, M. J. and Silvestri, S. (2010), *A comparative study of the representations of "suspect" communities in multi-ethnic Britain and their impact on Irish communities and Muslim communities—Mapping newspaper content*, London: London Metropolitan University, Institute for the Study of European Transformations, Working Paper 13

Nicolau, J. L. (2010), 'Testing prospect theory in airline demand', *Journal of Air Transport Management*, 17: 241–243

Nieuwenhuys, C. and Pécoud, A. (2007), 'Human trafficking, information campaigns, and strategies of migration control', *American Behavioral Scientist*, 50: 1674–1695

Noble, T. and Goffart, W. (2006), *From Roman provinces to medieval kingdoms*, New York: Routledge

Nonaka, I. and Takeuchi, H. (1995), *The knowledge creating company: How the Japanese companies create the dynamics of innovations*, New York: Oxford University Press

Nuissl, H. (2002), 'Bausteine des Vertrauens—Eine Begriffsanalyse', *Berliner Journal für Soziologie*, 1: 87–108

Nunn, N. and Qian, N. (2010), 'The Columbian exchange: A history of disease, food, and ideas', *Journal of Economic Perspectives*, 24: 163–188

O'Connell, P. G. J. (1997), 'Migration under uncertainty: "Try your luck" or "wait and see"', *Journal of Regional Science* 37: 331–347.

Ochs, E. and Capps, L. (1996), 'Narrating the self', *Annual Reviews Anthropology*, 25: 19–43

OECD (2010), *Open for business: Migrant entrepreneurship in OECD countries*, Paris: OECD

O'Malley, P. (2004), *Risk, uncertainty and government*, London: Cavendish Press/Glasshouse Press

Orozco, M. (2005), 'Hometown associations and development: Ownership, correspondence, sustainability and replicability', in Merz, B. J. (ed.), *New patterns for Mexico: Observations on remittances, philanthropic giving, and equitable development*, Cambridge, MA: Global Equity Initiative, Harvard University

Orozco, M. and Rouse, R. (2007), *Migrant hometown associations and opportunities for development: A global perspective*, Washington: Migration Policy Institute, Migration Information Source. Available at: www.migrationinformation.org/feature/display.cfm?ID=579 [Accessed October 2013]

Pahl, J. (1990), 'Household spending, personal spending and the control of money in marriage', *Sociology*, 24: 119–138

Pålsson, A. M. (1996), 'Does the degree of relative risk aversion vary with household characteristics?', *Journal of Economic Psychology*, 17: 771–787

Papademetriou, D. G. and Heuser, A. (2009), 'Council statement: Migration, public opinion and politics', in The Transatlantic Council on Migration (ed.), *Migration, public opinion and politics*, Gütersloh: Verlag Bertelsmann Stiftung, pp. 19–26

Paraskevis, D., Pybus, O., Magiorkinis, G., Hatzakis, A., Wensing, A. M. J. et al. (2009), 'Tracing the HIV-1 subtype B mobility in Europe: A phylogeographic approach', *Retrovirology*, 6: 49

Payne, J. W. (1976), 'Task complexity and contingent processing in decision making: An information search and protocol analysis', *Organizational Behavior and Human Performance*, 16: 366–387

Pebble Partnership (2012), *Outmigration and decreasing populations trends continue in southwest Alaska*. Available at: www.pebblepartnership.com/perch/resources/ebd-population-se0.pdf [Accessed October 2013]

Perch-Nielsen, S. L., Bättig, M. B. and Imboden, D. (2008), 'Exploring the link between climate change and migration', *Climatic Change*, 91: 375–393

Petros, M. (2005), *The costs of human smuggling and trafficking*, Global Migration Perspectives No. 31, Geneva: GCIM

Pidgeon, N., Kasperson, R. E. and Slovic, P. (2003), *The social amplification of risk*, Cambridge: Cambridge University Press

Piguet, E. (2008), *Climate change and forced migration: How can international policy respond to climate-induced displacement?*, Geneva: UNHCR Evaluation and Policy Analysis Unit

Piguet, E. (2010), 'Linking climate change, environmental degradation and migration: A methodological overview', *Wiley Interdisciplinary Reviews: Climate Change* 1: 517–524

Polanyi, M. (1966), *The tacit dimension*, London: Routledge & Kegan Paul

Poole, E. and Richardson, J. E. (eds.) (2006), *Muslims and the news media*, London and New York: I.B. Tauris

Portes, A. (1995), 'Economic sociology and the sociology of immigration: A conceptual overview', in Portes, A. (ed.), *The economic sociology of immigration: Essays on networks, ethnicity, and entrepreneurship*, New York: Russell Sage Foundation

Poston, D., Mao, X. M. and Yu, M. Y. (1994), 'The global distribution of the overseas Chinese around 1990', *Population and Development Review*, 20: 631–645

Power, J., Magnoni, B. and Zimmerman, E. (2011) *Formalizing the informal insurance inherent in migration: Exploring the potential links between migration, remittances and micro insurance*, Geneva: International Labour Organization, Microinsurance Paper No. 7

Pratt, J. W. (1964), 'Risk aversion in the small and in the large', *Econometrica*, 32: 122–136

Quétel, C. (1992), *History of syphilis* (1st ed.), Cambridge: Polity Press

Quiggin, J. (1982), 'A theory of anticipated utility', *Journal of Economic Behavior and Organization*, 3: 323–343

Quinn, T. C. (1994), 'Population migration and the spread of types 1 and 2 human immunodeficiency viruses', *Proceedings of the National Academy of Sciences of the United States of America*, 91: 2407–2414

Ratha, D. (2012), 'Remittances: Funds for the folks back home, *Finance and Development*, March 2012, Washington DC: World Bank

Renaud, F., Bogardi, J. J., Dun, O. and Warmer, K. (2007), *Control, adapt or flee: How to face environmental migration?*, InterSecTions No. 5, Bonn: United Nations University Institute for Environment and Human Security

Rhodes, T. (1997), 'Risk theory in epidemic times: Sex, drugs and the social organisation of "Risk behaviour"', *Sociology of Health & Illness*, 19: 208–227

Richardson, J. E. (2004), *(Mis)Representing Islam: The racism and rhetoric of British broadsheet newspapers*, Amsterdam: John Benjamins Publishing Company

Riley, P. (1988), 'Road culture of international long-term budget travelers', *Annals of Tourism Research*, 15: 313–328

Roberts, K. D. and Morris M. D. S. (2003), 'Fortune, risk, and remittances: An application of option theory to participation in village-based migration networks', *International Migration Review*, 37: 1252–1281

Rogaly, B. (2008), 'Migrant worker in the ILO's global alliance against forced labour report: A critical appraisal', *Third World Quarterly*, 29: 1431–1447

Rogaly, B. and Rafique, A. (2003), 'Struggling to save cash: Seasonal migration and vulnerability in West Bengal, India', *Development and Change*, 34: 659–681

Ronay, R. and Kim, D.Y. (2006), 'Gender differences in explicit and implicit risk attitudes: A socially facilitated phenomenon', *British Journal of Social Psychology*, 45: 397–419

Rosa, E. A. (2003), 'The logical structure of the social amplification of risk framework (SARF): Metatheoretical foundation and policy implications', in Pidegeon, N.,

Kaspersen, R. E. and Slovic, P. (eds.), *The social amplification of risk*, Cambridge: Cambridge University Press, pp. 47–76

Rose, N. (1990), *Governing the soul: The shaping of the private self*, Routledge: London

Rose, N. (1996a), 'The death of the social? Refiguring the territory of government', *Economy and Society*, 25: 327–356

Rose, N. (1996b), 'Identity, genealogy, history', in Hall, S. and Du Gay, P. (eds.), *Questions of cultural identity*, London: Sage, pp. 128–150

Rose, N. (1999), *Powers of freedom*, Cambridge: Cambridge University Press

Rosenau, J. N. (1997), *Along the domestic-foreign frontier: Exploring governance in a turbulent world*, Cambridge: Cambridge University Press

Rosenzweig, M. R. and Stark, O. (1989), 'Consumption smoothing, migration and marriage: Evidence from rural India', *Journal of Political Economy*, 97: 905–926

Rothschild, B. M. (2005), 'History of syphilis', *Clinical Infectious Diseases*, 40: 1454–1463

Rozelle, S., Taylor, J. E. and de Brauw, A. (1999), 'Migration, remittances, and agricultural productivity in China', *American Economic Review*, 89: 287–291

Rudolph, C. (2003), 'Security and the political economy of international migration', *American Political Science Review*, 97: 603–620

Rudnyckyj, D. (2004), 'Technologies of servitude: Governmentality and Indonesian transnational labor migration', *Anthropological Quarterly*, 77: 407–434

Saarela, J. and Rooth, D. O. (2012), 'Uncertainty and international return migration: Some evidence from linked register data', *Applied Economics Letters*, 19: 1893–1897

Sacerdote, B. (2002), 'The nature and nurture of economic outcomes', *American Economic Review*, 92: 344–348

Saggar, S. (2008), *Follow my lead: The role of politics in shaping the debate on labour migration*, London: Policy Network Briefing

Sahm, C. R. (2007), *Does risk tolerance change?*, Ann Arbor: University of Michigan, Job Market Papers

Salt, J. and Stein, J. (1997), 'Migration as a business: The case of trafficking', *International Migration*, 35: 467–494

Sanbonmatsu, D. M., Kardes, F. R and Herr, P. M. (1992), 'The role of prior knowledge and missing information in multiattribute evaluation', *Organizational Behavior & Human Decision Processes* 51: 76–91

Saran, A. (2009), 'A commentary on public attitudes on immigration: The United Kingdom in an international context', in The Transatlantic Council on Migration (ed.), *Migration, public opinion and politics*, Gütersloh: Verlag Bertelsmann Stiftung, pp. 155–166

Sassen, S. (1996), *Losing control? Sovereignty in an age of globalization*, New York: Columbia University Press

Sassen, S. (2004), 'Beyond sovereignty: *De-facto* transnationalism in immigration policy', in Friedman, J. and Randeria, S. (eds.), *Worlds on the : Globalisation, migration and cultural security*, Toda Institute book series on global peace and policy, London: I.B. Tauris

Savage, L. J. (1954), *The foundations of statistics*, originally published by New York: Wiley, reprinted by the Dover Publications (2nd Revised edition, 1972)

Schaffer, E. (2008), 'Remittances and reputations in hawala money-transfer systems: Self-enforcing exchange on an international scale', *The Journal of Private Enterprise*, 24: 1–17

Schooley, D. K. and Worden, D. D. (1996), 'Risk aversion measures: Comparing attitudes and asset allocation', *Financial Services Review*, 5: 87–99

Schwartz, B., Ben-Haim, Y. and Dacso, C. (2011), 'What makes a good decision? Robust satisficing as a normative standard of rational behaviour', *The Journal for the Theory of Social Behaviour*, 41: 209–227

Schwartz, M. and Thompson, M. (1990), *Divided we stand: Redefining politics, technology and social choice*, Hemel Hempstead: Harvester Wheatsheaf

Scott, J. C. (1998), *Seeing like a state: How certain schemes to improve the human condition have failed*, New Haven, CT: Yale University Press

Shah, A. K. and Oppenheimer, D. M. (2008), 'Heuristics made easy: An effort-reduction framework', *Psychological Bulletin*, 134: 207–222

Sharland, E. (2006), 'Young people, risk taking, and risk making: Some thoughts for social work', Forum Qualitative Sozialforschung / Forum: Qualitative Social Research, 7, Art. 23, Available at www.qualitative-research.net/index.php/fqs/article/view/56. [Accessed October 2013]

Shepherd, C. D. and Woodruff, R. B. (1988), 'A muddling through model of family purchase conflict management', in *Proceedings of the Society for Consumer Psychology*, Society for Consumer Psychology, pp. 73–86

Sherbinin, A. de, Castro, M., Gemenne, F., Cernea, M. M., Adamo, S., Fearnside, P. M., Krieger, G., Lahmani, S., Oliver-Smith, A., Pankhurst, A., Scudder, T., Singer, B., Tan, Y., Wannier, G., Boncour, P., Ehrhart, C., Hugo, G., Pandey, B. and Shi, G. (2011), 'Preparing for resettlement associated with climate change', *Science*, 334: 456–57

Shields, R. (1991), *Places on the margin: Alternative geographies of modernity*, London: Routledge

Siegrist, M. and Cvetkovich, G. (2000), 'Perception of hazards: The role of social trust and knowledge', *Risk Analysis*, 20: 713–719

Simon, H. A. (1955), 'A behavioral model of rational choice', *The Quarterly Journal of Economics*, 69: 99–118

Simon, H. A. (1986), 'Rationality in psychology and economics', *Journal of Business* (Supplement), 59: S209-S224

Simmons, C. J. and Lynch, J. G. (1991), 'Inference effects without inference making? Effects of missing information on discounting and use of presented information', *Journal of Consumer Research*, 17: 477–491

Sjastaad, L. A. (1962), 'The costs and returns of human migration', *Journal of Political Economy*, 70S: 80–93

Slovic, P., Fischhoff, B. and Lichtenstein, S. (1979), 'Rating the risks', *Environment*, 21: 14–39

Smith, A. (1759), *Theory of moral sentiments*, 2005 reprint of 1790 London sixth edition, Sao Paulo, Brasil: MetaLibri

Smith, T. R. and Clark, W. A. V. (1979), 'Decision-making and search model for intraurban migration', *Geographical*, 11: 1–22

Solomon, M. R., Bamossy, G., Askegaard, S., & Hogg, M. K. (2006). *Consumer Behaviour: A European Perspective* (5th ed.). London: Financial Times/Prentice Hall

SOSR, Statistical Office of the Slovak Republic (2012), Labour force survey. Available at: portal.statistics.sk/showdoc.do?docid=23558 [Accessed 18 April 2013]

Soto, S. M. (2009), 'Human migration and infectious diseases', *Clinical Microbiology and Infection*, 15: 26–28

Stark, O. (1991), *The migration of labour*, Oxford: Blackwell

Stark, O. and Bloom, D. E. (1985), 'The new economics of labour migration', *American Economic Review*, 75: 173–178

Stark, O. and Levhari, D. (1982), 'On migration and risk in LDCs', *Economic Development and Cultural Change*, 31: 191–196

Stark, O. and Lucas, R. E. B. (1988), 'Migration, remittances and the family', *Economic development and cultural change*, 36: 465–481

Steinbeck, J. (1939), *The grapes of wrath,* London: Heinemann

Stern, N. (2007), *The economics of climate change: The Stern review*, Cambridge: Cambridge University Press

Stigler, G. (1961), 'Economics of information', *The Journal of Political Economy*, 69: 213–225

Stiglitz, J. E. (2000), 'The contribution of the economics of information to twentieth century economics', *The Quarterly Journal of Economics*, 115: 1441–1478

Stott, H. P. (2006), 'Cumulative prospect theory's functional menagerie', *Journal of Risk and Uncertainty*, 32: 101–130

Styhre, A. (2004), 'Rethinking knowledge: A Bergsonian critique of the notion of tacit knowledge', *British Journal of Management*, 15: 177–188

Suleri, A. and Savage, K. (2006), *Remittances in crisis: A case study from Pakistan*, Humanitarian Policy Group, London: Overseas Development Institute

Sun, S. and Manson, M. (2010), 'An agent-based model of housing search and intraurban migration in the Twin Cities of Minnesota', in David A. Swayne (chair), *International Congress on Environmental Modelling and Software Modelling for Environment's Sake, Fifth Biennial Meeting*, Symposium conducted at the meeting of the International Environmental Modelling and Software Society (iEMSs), Ottawa, Canada. Available at www.iemss.org/iemss2010/index.php?n=Main.Proceedings [Accessed October 2013]

Sung, J. and Hanna, S. (1996), 'Factors related to risk tolerance', *Financial Counselling and Planning*, 7: 11–20

Suro, R. (2009), 'Promoting stalemate: The media and US policy on migration', in The Transatlantic Council on Migration (ed.), *Migration, public opinion and politics*, Gütersloh: Verlag Bertelsmann Stiftung, pp. 185–221

Tacoli, C. (2011), *Not only climate change: Mobility, vulnerability and socioeconomic transformation in environmentally-fragile areas of Bolivia, Senegal and Tanzania*, International Institute for Environment and Development, London. Available at: pubs.iied.org/10590IIED.html [Accessed June 2012]

Tainter, J. (1988), '*The collapse of complex societies*', Cambridge: Cambridge University Press

Taylor, J. E. (1992), 'Remittances and inequality reconsidered: Direct, indirect, and intertemporal effects', *Journal of Policy Modeling*, 14: 187–208

Taylor, J. E. (2006), *International migration and development*, International Symposium on International Migration and Development, Population Division, United Nations Secretariat, Rurin, 28–30 June 2006

Taylor, J. E. and López-Feldman, A. (2007), *Does migration make rural households more productive? Evidence from Mexico*. Food and Agriculture Organization of the United Nations, Agricultural Development Economics Division, ESA Working Paper No. 07–10

Taylor, J. E., Rozelle, S. and de Brauw, A. (2003), 'Migration and incomes in source communities: A new economics of migration perspective from China', *Economic Development and Cultural Change*, 52: 75–102

Taylor-Gooby P. and Zinn, J. O. (2005), *Changing directions in risk research: Reinvigorating the social*, Canterbury: University of Kent, SCARR Working Paper 2005/8

Tenner, E. (1997), *Why things bite back: Technology and the revenge of unintended consequences*, New York: Alfred A. Knopf

Thompson, M. (1980), 'The aesthetics of risk: culture or conflict', in Schwing, R. S. and Albers, W. A. (eds.), *Societal risk assessment: How safe is safe enough?*, New York: Plenum, pp. 273–285

Thompson, M, Ellis, R. and Wildavsky, A. (1990), *Cultural theory*, Boulder, Colorado: Westview

Thornton, R. (1987), *American Indian holocaust and survival: A population history since 1492*, Norman: University of Oklahoma Press, Publishing Division of the University

Threadgold, T. (2009), 'The media and migration in the United Kingdom, 1999 to 2009', in The Transatlantic Council on Migration (ed.), *Migration, public opinion and politics*, Gütersloh: Verlag Bertelsmann Stiftung, pp. 222–260

Tierney, K. J. (1999), 'Toward a critical sociology of risk', *Sociological Forum*, 14: 215–242

Todaro, M. P. (1969), 'A model of labor migration and urban unemployment in less developed countries', *The American Economic Review*, 59: 138–148

Todaro, M. P. (1980), 'International migration in developing counties: A survey', (Chapter 6) in Easterlin, R. A. (ed.), *Population and economic change in developing countries*, Chicago: NBER, pp. 361–402

Todaro, M. P. and Maruszko, L. (1987), 'Illegal migration and US immigration reform: A conceptual framework', *Population and Development Review*, 13: 101–114

Torpey, J. (1999), *The invention of the passport: Surveillance, citizenship, and the state*, Cambridge: Cambridge University Press

Tulloch, J. and Lupton, D. (2003), *Risk and everyday life*, London: Sage

Tunali, I. (2000), 'Rationality of migration', *International Economic Review*, 41: 893–920

Turner, B. (2001), 'Risks, rights and regulation: An overview', Health, Risk and Society, 3: 9–18

Tversky, A. (1972), 'Elimination by aspects: A theory of choice', *Psychological Review*, 79: 281–299

Tversky, A. and Fox, C. (1995), 'Weighing risk and uncertainty', *Psychological Review*, 102: 269–283

Tversky, A. and Kahneman, D. (1974), 'Judgement under uncertainty: Heuristics and biases', *Science*, 185: 1124–1131

Tversky, A. and Kahneman, D. (1992), 'Advances in prospect theory: Cumulative representation of uncertainty', *Journal of Risk and Uncertainty*, 5: 297–323

Tversky, A. and Koehler, D. J. (1994), 'Support theory: A nonextentional representation of subjective probability', *Psychological Review*, 101: 547–567

Umblijs, J. (2012), *The effect of networks and risk attitudes on the dynamics of migration*, Oxford: International Migration Institute, University of Oxford, Working Paper 54

United Nations (2012), *World urbanization prospects, the 2011 Revision*. Available at: esa.un.org/unup/CD-ROM/Urban-Agglomerations.htm

UNODC (2010) *A short introduction to migrant smuggling*, Issue paper, United Nations, Office on Drugs and Crime. Available at: www.unodc.org/documents/human-trafficking/Migrant-Smuggling/Issue-Papers/Issue_Paper_-_A_short_introduction_to_migrant_smuggling.pdf

Urry, J. (2007), *Mobilities*, Cambridge: Polity

Urry, J. (2012), 'Social networks, mobile lives and social inequalities', *Journal of Transport Geography* 21: 24–30

Vaughan, E. and Dunton, G. F. (2007), 'Difficult socio-economic circumstances and the utilization of risk information: A study of Mexican agricultural workers in the USA', *Health, Risk & Society*, 9: 323–341

Väyrynen, R. (2003), *Illegal immigration, human trafficking, and organized crime*, Helsinki: United Nations University/WIDER

Verplanken, B., Hazenberg, P. T. and Palenéwen, G. R. (1992), 'Need for cognition and external information search effort', *Journal of Research in Personality*, 26: 128–136

Vertovec, S. (2002), *Transnational networks and skilled labour migration*, ESRC Transnational Communities Programme, WPTC 02 02, Oxford

Vogel, D. (2001), *The new politics of risk regulation in Europe*, London: London School of Economics, Centre for Analysis of Risk and Regulation, Discussion Paper 2003/72, Helsinki: United Nations University/WIDER

Voluntary Health Association of India (2000), *Report on environmental sanitation and community water supply: Country situation analysis*, New Delhi

Vullnetari, J. (2012), *Albania on the move: Links between internal and international migration*, Amsterdam: Amsterdam University Press

Waldinger, R., Aldrich, H. and Ward, R. (eds.) (1990), 'Opportunities, group characteristics, and strategies', in Waldinger, R., Aldrich, H. and Ward, R. (eds.), *Ethnic Entrepreneurs*, Sage: Newbury Park, pp. 13–48

Walters, W. (2004), 'The political rationality of European integration', in Larner, W. and Walters, W. (eds.), *Global governmentaliy: Governing international spaces*, New York: Routledge, pp. 155–173

Walters, W. (2008), 'Putting the migration-security complex in its place', in Amoore, L. and de Goede, M. (eds.), *Risk and the war on terror*, London: Routledge, pp. 158–177

Wang, Y. (1995), 'A study on the migration policy in ancient China', *China Journal of Population Science*, 7: 27–38

Warren, C., McGraw, A. P. and van Boven, L. (2011), 'Values and preferences: defining preference construction', *Wiley Interdisciplinary Reviews: Cognitive Science* 2: 193–205

Weber, E. U. and Hsee, C. K. (1998), 'Cross-cultural differences in risk perception, but cross-cultural similarities in attitude towards perceived risk', *Management Science*, 44: 1205–1217

Weiner, M. (1995), *The global migration crisis: Challenge to States and to Human Rights*, New York: HarperCollins

Weiner M. (1996), *International migration and security*, Boulder, CO: Westview Press

White, C. M. and Koehler, D. J. (2004), 'Missing information in multiple-cue probability learning', *Memory & Cognition*, 32: 1007–1018

WHO, World Health Organisation (2007), *Tuberculosis and migration*, WHO: 3 September 2007. Available at: www.euro.who.int/en/what-we-do/health-topics/communicable-diseases/tuberculosis/publications/pre-2009/tuberculosis-and-migration [Accessed June 2012]

WHO, World Health Organisation (2013a), *Sexually transmitted infections, WHO/ RHR/13.02 2013* Available at: apps.who.int/iris/bitstream/10665/82207/1/WHO_RHR_13.02_eng.pdf

WHO, World Health Organisation (2013b): *Global alert and response*. Available at: www.who.int/csr/alertresponse/epidemicintelligence/en/ [Accessed May 2014]

Wildavsky, A. (1991), 'If claims of harm from technology are false, mostly false, or unproven, what does that tell us about science?', in Berger, P. et al. (eds.), *Health, lifestyle and environment*, London: Social Affairs Unit/Manhattan Institute

Williams, A. M. (2007), 'International labour migration and tacit knowledge transactions: A multi-level perspective', *Global Networks*, 7: 29–50

Williams, A. M. (2008), 'Employability and international migration: Theoretical perspectives', in MacKay, S. (ed.), *Refugees, Recent Migrants and Employment: Challenging barriers and exploring pathways*', Oxford: Routledge, pp. 23–34

Williams, A. M. (2013), 'Mobilities and sutainable tourism: Path-dependent or path-creating relationships?', *Journal of Sustainable Tourism*, 21: 511–513

Williams, A. M. and Baláž, V. (2004), 'From private to public sphere, the commodification of the au pair experience? Returned migrants from Slovakia to the UK', *Environment and Planning A*, 36: 1813–1833

Williams, A. M. and Baláž, V. (2005), 'What human capital, which migrants? Returned skilled migration to Slovakia from the UK', *International Migration Review*, 39: 438–469

Williams, A. M. and Baláž, V. (2008), *International migration and knowledge*, Routledge: London

Williams, A. M. and Baláž, V. (2009), 'Low-cost carriers, economies of flows and regional externalities', *Regional Studies*, 43: 667–691

Williams, A. M. and Baláž, V. (2012), 'Migration, risk and uncertainty: Theoretical perspectives', *Population, Space and Place*, 18: 167–180

Williams, A. M. and Baláž, V. (2013), 'Mobility, risk tolerance and competence to manage risks', forthcoming in: *Journal of Risk Research*, published online first, doi: 10.1080/13669877.2013.841729

Williams A.M. and Baláž, V. (2014), 'Tourism, risk and uncertainty: Theoretical reflections', forthcoming in: *Journal of Travel Research*, published online first, doi:10.1177/0047287514523334

Williams A. M., Chaban N. and Holland M. (2011), 'The circular international migration of New Zealanders: Enfolded mobilities and relational places', Mobilities, 6: 125–147.

Williams, P. and Soutar, G. (2009), 'Value, satisfaction and behavioural intentions in an adventure tourism context', *Annals of Tourism Research* [online], 36: 416–438

Willis, H. H. (2007), 'Guiding resource allocations based on terrorism risk', *Risk Analysis*, 27: 597–606

Winters, P., de Janvry, A. and Sadoulet, E. (2001), 'Family and community networks in Mexico-U.S. migration', *The Journal of Human Resources*, 36: 159–184

Wirth, T., Hildebrand, F., Allix-Béguec, C., Wölbeling, F., Kubica, T. et al. (2008), 'Origin, Spread and Demography of the Mycobacterium tuberculosis Complex', *PLoS Pathogens* 4: e1000160 doi:10.1371/journal.ppat.1000160

Wolf, D. (1990), 'Daughters, decisions and domination: An empirical and conceptual critique of household strategies', *Development and Change*, 21: 43–74

Wolpert, J. (1965), 'Behavioural aspects of the decision to migrate', *Papers of the Regional Science Association*, 15: 159–169

Wood, M. (2001), *In the footsteps of Alexander the Great: A journey from Greece to Asia*, Berkeley: University of California Press

World Bank (2012) Databases on GDP and remittances. Available at: web.worldbank.org/WBSITE/EXTERNAL/TOPICS/0,,contentMDK:21924020~pagePK:5105988~piPK:360975~theSitePK:214971,00.html [Accessed December 2013]

WTO, World Tourism Organization (2013): *Tourism highlights, 2013 edition*, Geneva: World Tourism Organisation

Wu, G. and Gonzalez, R. (1996), 'Curvature of the probability weighting function', *Management Science*, 42: 1676–1690

Yao, R., Hanna, S. D. and Lindamood, S. (2004), 'Changes in financial risk tolerance, 1983–2001', *Financial Services Review*, 13: 249–266

Zadeh, L. A. (1965), 'Fuzzy sets', *Information and Control*, 8: 338–353

Zambonini, G. (2009), 'The evolution of German media coverage of migration', in The Transatlantic Council on Migration (ed.), *Migration, public opinion and politics*, Gütersloh: Verlag Bertelsmann Stiftung, pp. 169–184

Zhang, W. and Semmler, W. (2009), 'Prospect theory for stock markets: Empirical evidence with time-series data', *Journal of Economic Behavior & Organization*, 72: 835–849

Zimmerman, C., Kiss, L., and Hossain, M. (2011), 'Migration and health: A framework for 21st century policy-making', *PLoS Med* 8: e1001034 doi:10.1371/journal.pmed.1001034

Zinn, J. O. (2004a), *Literature review: Sociology and risk*, Canterbury: University of Kent, SCARR Working Paper 2004/1

Zinn J. (2004b), *Literature review: Economics and risk*, Canterbury: University of Kent, SCARR Working Paper 2004/2

Zinn, J. (2007), *Risk, social change and morals: Conceptual approaches of sociological risk theories*, Canterbury: University of Kent, SCARR Working Paper 2007/17

Zinn, J. O. (2008), 'Introduction: The contribution of sociology to the discourse on risk and uncertainty' in Taylor-Gooby, P., and Zinn, J. O. (eds.), *Risk in social science*, Oxford: Oxford University Press, pp. 1–17

Zinn, J. O. and Taylor-Gooby, P. (2006a), 'Risk as an interdisciplinary research area', in Taylor-Gooby, P. and Zinn, J. O. (eds.), *Risk in social science*, Oxford: Oxford University Press, pp. 20–53

Zinn, J. O. and Taylor-Gooby, P. (2006b), 'The challenge of (managing) new risk', in Taylor-Gooby, P. and Zinn, J. O. (eds.), *Risk in social science*, Oxford: Oxford University Press, pp. 54–76

van Zuuren, F. J. and Wolfs, H. M. (1991), 'Styles of information seeking under threat: Personal and situational aspects of monitoring and blunting', *Personality and Individual Differences*, 12: 141–149

Zyphur, M., Narayanan, J., Arvey, R. D. and Alexander, G. J. (2009), 'The genetics of economic risk preferences', *Journal of Behavioral Decision Making*, 22: 367–377

Index

For Product Safety Concerns and Information please contact our EU
representative GPSR@taylorandfrancis.com
Taylor & Francis Verlag GmbH, Kaufingerstraße 24, 80331 München, Germany